Trees

A Complete Guide to Their Biology and Structure

Roland Ennos

Comstock Publishing Associates

a division of
Cornell University Press
Ithaca, New York

First published by the Natural History Museum, Cromwell Road, London SW7 5BD
© The Trustees of the Natural History Museum, London 2001
This edition © The Trustees of the Natural History Museum, London 2016

First published in the United States of America 2016 by Cornell University Press
First printing, Cornell Paperbacks, 2016

Library of Congress Cataloging-in-Publication Data
Names: Ennos, A. R., author.
Title: Trees : a complete guide to their biology and structure / Roland Ennos.
Description: Ithaca, New York : Comstock Publishing Associates, a division of
    Cornell University Press, 2016. | "First published by the Natural History
    Museum ... London 2001"--Title page verso. | Includes bibliographical
    references and index.
Identifiers: LCCN 2016003422 | ISBN 9781501704932 (pbk. : alk. paper)
Subjects: LCSH: Trees.
Classification: LCC QK475 .E552 2016 | DDC 582.16--dc23
LC record available at http://lccn.loc.gov/2016003422

Designed by Mercer Design, London
Reproduction by Saxon Digital Services
Printed by C&C Offset Printing Co, Ltd.

Front cover: Pine forest, Dalarna, Sweden ©Janos Jurka/The Trustees of the Natural History Museum, London

Paperback printing 10 9 8 7 6 5 4 3 2 1

Contents

	Introduction	4
1	The tree story	8
2	How trees lift water	28
3	How trees stand up	36
4	Limits to height	52
5	Survival strategies	58
6	Trees in different climates	68
7	Specialist trees	86
8	Southern hemisphere trees	94
9	Trees and people	102
	Glossary	124
	Further information	125
	Index	126
	Picture credits	128

Introduction

TREES ARE PERHAPS THE MOST familiar of all organisms. They grow all around us, even in towns, and they are easier to study than animals because they cannot run away. Trees are also economically valuable, and we still use more wood than any other engineering material. Most of us can even recognize a number of species. So it is tempting to think that we know all about them.

However, there is much more to trees than you might think. Not only are they the largest organisms that have ever lived – some giant redwoods are 10 times heavier than a full-grown blue whale – but they have also dominated the dry land for over 300 million years – far longer than the dinosaurs or mammals. Trees are also extremely diverse, with many thousands of species, which live in a wide range of habitats. What is more, when we ask ourselves some simple questions we soon reveal that we know very little, even about the trees that are growing outside our window. Why do they only have one trunk, for instance? Why are there so many species? And why are some of them evergreen while others are deciduous?

This book investigates the world of trees by asking just these sorts of questions. As we shall see, the key to understanding trees is to treat them as living organisms, struggling for survival in a hostile world. Looking at them in this light we can start to appreciate not only their beauty, but also the sheer ingenuity of their structures and lifestyles.

There is no doubt that trees are extremely successful. There are more than 80,000 species – which range from tiny arctic willows a few centimetres high to giant redwoods that can grow over 100 metres (328 feet) tall – and forests cover over 30 per cent of the dry land. The first question we must ask is why do trees do so well? The main difficulty in answering this question is that trees are not what biologists call a 'natural group'. Whereas other successful groups of plants, such as grasses, all descend from a single common ancestor, the tree form has evolved many times in the course of history; the ancestors of different types of tree were quite unrelated plants. If you look closely at trees, this is very obvious because, unlike grasses, they may sport a bewildering variety of reproductive structures. The fact that oaks produce acorns while pines produce cones, for example, tells you that they are only very distantly related to each other. So although a tree superficially looks like its tree neighbours,

OPPOSITE Trees compete for the light by holding their canopy of leaves high above the ground, as seen here in this birch forest, Saint John, New Brunswick, Canada.

it may in fact be more closely related to a small herb or shrub. The giant *Koompassia excelsa* from the rainforests of Borneo is closely related to tiny clover plants. The giveaway is that both tree and clover produce flowers that look like sweet peas!

It is clear, therefore, that the success of trees is not due to their reproductive characteristics; it must instead be due to their basic body plan. The accepted definition of a tree is that it is a plant with a more or less permanent shoot system that is supported by a single woody trunk. At first sight this might not seem to be a good design, because only a small fraction of a tree's biomass – its leaves – is capable of producing the sugars it needs to grow by the process of photosynthesis. Therefore much of the energy produced by the leaves has to be diverted to make unproductive tissue (such as the woody trunk, branches and roots) as the tree grows. In comparison, a plant composed almost entirely of a flat plate of photosynthetic cells (like duckweed) has the capability of growing and reproducing much faster. It must be remembered, however, that plants do not grow on their own but must compete with other plants for water, nutrients – and especially light.

Trees are, indeed, excellent competitors for light; they are able to hold their leaves above those of other plants and shade them out. One reason they can do this is that they have a permanent structure above the ground. Each year the shoots of a tree have a head start over those of non-woody herbaceous plants; they grow from buds high up on the tree, whereas the herbaceous shoots, which die back each year, have to grow up from the ground. Of course shrubs also have a permanent woody structure, so how do trees outcompete shrubs for light? The answer lies in the differences in their structure. Shrubs typically branch near the ground and so have several narrow stems rather than a single trunk. These stems can support lots of leaves, but weight for weight they are less rigid than a single thick trunk. For this reason, shrubs do not typically grow as tall as trees; the stems bow down under their own weight, whereas the single rigid trunk of a tree is easily able to support itself. The best strategy for a tall plant is to grow with a single trunk that branches near the top. Even so, this structural design has a downside; the woody trunk and branches are energetically expensive to construct. This makes trees relatively slow-growing compared with herbaceous plants. However, trees are able to live a long time because wood is a strong, long-lasting material; it is also cheap to maintain because most of the cells in it are dead.

If a patch of soil is cleared, it will initially be colonized by fast-growing herbs and shrubs. The trees will grow slowly, but they will keep on growing until they are taller than the herbs and shrubs. Eventually the trees will cast such dense shade that most of the other plants, bar a few woodland herbs, will die out and the area will turn into a forest. One or two hundred years after the clearance, it will look as if a forest had always been there. Trees have such an advantage in the long term, therefore, that it is not surprising that the tree habit has evolved so often.

However, just as there is more than one way of building a bridge, so there is more than one way of growing into a tree. As we shall see in the next chapter, different groups of land plants have evolved into trees, and though superficially similar, they are very different under their bark.

BELOW An oak tree, *Quercus robur*, in spring, showing the characteristic structure of a typical tree: a single trunk supporting a branching crown.

CHAPTER 1

The tree story

PLANTS FIRST COLONIZED LAND DURING the Ordovician period, around 470 million years ago. The first land plants were almost certainly plate-like organisms that were made up of a single layer of cells, rather like some modern liverworts. In many ways these plants were ideally suited to the new habitat. They lived in the thin film of water on the surface of the ground, so they had no problems with obtaining water, and all their cells were photosynthetic, so the plants could grow rapidly. The only problem was that they were unable to grow up away from the surface: flat plates are just too flexible to support themselves. They were therefore poor competitors for light and would have been very vulnerable to being shaded by any plant that had developed a rigid supporting structure.

THE FIRST VASCULAR PLANTS

The first step in the evolution of trees was to produce an upright shoot system. The earliest free-standing plants evolved in the Silurian period and had shoot systems that looked like simple branched cylinders. The best known of these early plants was *Rhynia*, which thrived 420 million years ago around the margins of hot springs. Its shoots grew upwards from underground stems, or rhizomes. They were waterproofed by a cuticle that covered their surface and they supported themselves by turgor. Turgor is the name given to the process whereby the fleshy outer cells of plant stems are stiffened by being filled with water under pressure, just as inflatable dinghies are stiffened by pressurized air; it can be seen in the stalks of modern-day plants such as dandelions. *Rhynia* obtained carbon dioxide for its photosynthesis through holes in its surface called stomata. Unfortunately, water was lost by the same route. This water was replaced by absorption through tiny underground rhizoid cells, which grew out of the rhizome. The water was then transported up the shoots through a central strand of so-called vascular tissue.

ABOVE Reconstruction of the early vascular plant *Rhynia* from the Silurian period, showing the branched aerial stems and underground rhizomes.

OPPOSITE Modern tree ferns, *Cyathea* sp., growing in the understory of a modern rainforest.

The main component of the vascular tissue, xylem, was a major key to the success of plants like *Rbynia*. Xylem conducted water well because it was made up of rows of dead cells joined together end-to-end to form pipes. A small strand of xylem could therefore supply the whole stem with the water it needed. The walls of the xylem cells were strengthened and made 'woody' by incorporating a chemical called lignin. This prevented the cells from collapsing and it also meant that the xylem tissue could help support the stem.

THE FIRST TREES

Rhynia was well suited to a life on bare saltwater margins; indeed it closely resembled modern saltmarsh plants such as the glasswort, *Salicornia*. However, both these plants have several characteristics that would lead them to be outcompeted for light in areas that were more crowded with plants. *Rhynia* did not have any leaves so would have captured little light and consequently grown fairly slowly. Its fleshy stems were weak, which limited it to a height of around 20 centimetres (8 inches).

It also did not have proper roots, and this too would have limited its ability to anchor a larger shoot system or supply it with water. In better rooting media, keener competition for light and space drove three very different groups of plants to evolve adaptations that allowed them to overcome these problems: they developed thick trunks, anchoring roots, and flat, leaf-like appendages. These plants became the first trees.

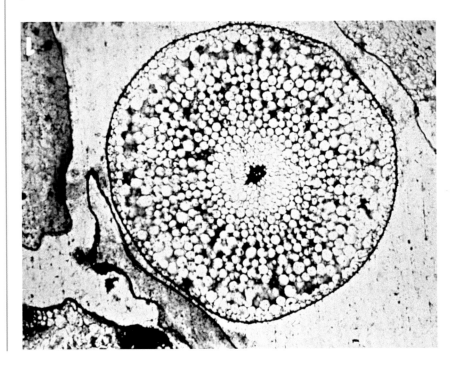

RIGHT Cross section through the stem of *Rhynia*, showing the tiny central strand of xylem cells, which transported water, and the larger outer cells, which supported the plant by turgor.

CLUB MOSSES

The first trees were the lycophytes, or club mosses. They improved their ability to capture light by developing simple scale-like leaves; these covered the branches and each was supplied with water by a single strand of xylem. The club mosses managed to improve their water-conducting ability as they grew by thickening the central strand of xylem. This allowed even tall trees to transport enough water up their stem to their crown. However, because the xylem was located in the centre of the stem, this did not make the stem significantly stronger. Instead, club mosses strengthened their stems by encasing them in a cylinder of bark-like material and added more of this material as the plant grew. Club mosses were therefore the first plants to produce a permanent trunk that could grow thicker as it grew taller. They anchored their trunks in the ground with special horizontal branches called rhizophores that superficially resembled the lateral roots of modern trees. They developed bare, root-like organs that radiated from the rhizophores, taking up water and anchoring the trunk, just like the sinker roots. Thus, though the club mosses grew in a very different way from modern trees, similar selection pressures made them look surprisingly similar. This is a common phenomenon in biology and is known as convergent evolution.

BELOW LEFT Cross section through the trunk of *Lepidodendron*, showing the central water-conducting xylem and the outer layer of supporting bark-like tissue.

BELOW Reconstruction of the giant Carboniferous club moss *Lepidodendron*, which grew up to 40 metres (131 feet) tall.

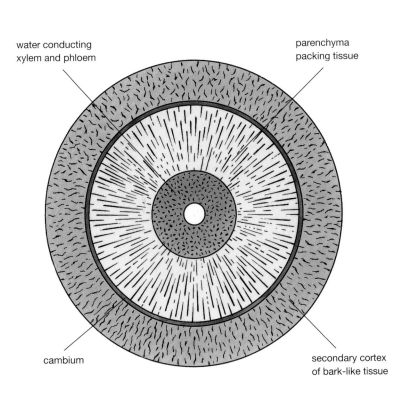

water conducting xylem and phloem

parenchyma packing tissue

cambium

secondary cortex of bark-like tissue

ABOVE The 330-million-year-old fossil stumps of *Lepidodendron* trees from the fossil grove, Victoria Park, Glasgow, UK.

The club mosses reached a truly gigantic size by the Carboniferous period. One of the best-known species, *Lepidodendron*, grew up to 40 metres (131 feet) tall. With its straight trunk, small, branched crown and spear-like leaves, it must have looked something like a modern monkey puzzle tree. Many fossils of *Lepidodendron* have been found. The most exciting are probably the 330-million-year-old tree stumps at the Forest Grove in Victoria Park, Glasgow, UK. They stand in the original positions where they died and are an impressive reminder of an exotic past.

HORSETAILS

A second group of plants, the pterophytes, developed two very different ways of producing a thick, strong trunk. One group of pterophytes – the horsetails – arranged their xylem not in a central strand, but in a ring around the outside of the stem; this meant that the trunk was hollow, but it was strengthened along its length by regularly spaced bulkheads, which separated the central hollow into compartments. A similar arrangement can be seen in the stems of bamboo, and the trunks of *Cecropia* (see p. 35). The trunk grew by extending at these bulkheads, and all the branches radiated out from them. This pattern of growth produced a shoot system that looked rather like a giant bottle brush. The trunks grew from large underground stems or rhizomes, just as in the more primitive *Rhynia*; however, the horsetails developed true roots that anchored the shoots and supplied them with water. By the Carboniferous period, horsetails had managed to grow into giant trees, such as the 30-metre (98-foot)-tall *Calamites*. It had no leaves, but its stem branched several times to produce many photosynthetic twigs, which looked not unlike the needles of pine trees.

BELOW LEFT Reconstruction of the Carboniferous horsetail *Calamites*, showing the regular pattern of branching from the jointed trunk, and the underground rhizome and roots.

BELOW Cross section through the hollow trunk of *Calamites*, showing the outer ring of xylem that both transported water and supported the plant.

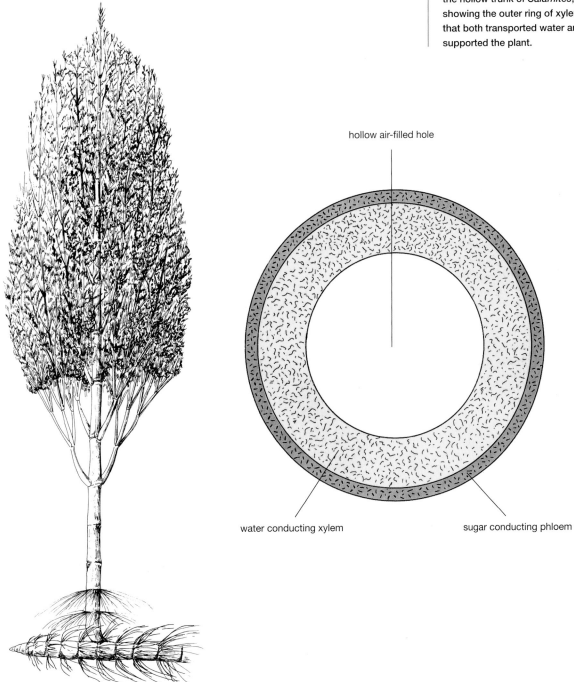

hollow air-filled hole

water conducting xylem

sugar conducting phloem

FERNS

Other pterophytes that had members which evolved into trees were the ferns. They developed more luxuriant foliage than either the club mosses or the horsetails. The leaves developed as a photosynthetic membrane, 'filling in' the area between flattened sprays of twigs. However, ferns never developed such sophisticated methods of producing a trunk as the other two groups of plants. They only have narrow stems that are not strong enough on their own to support the leaves.

Tree ferns, however, make a rigid trunk by surrounding the stem with the bases of shed leaves, and binding up the structure with aerial roots. This sort of trunk is strong but cannot branch, so tree ferns cannot produce a complex crown. Even today, tree ferns simply look like ordinary shuttlecock ferns mounted on pedestals. Despite these disadvantages, however, the tree fern *Psaronius*, which flourished in the Carboniferous period, still managed to reach a height of 20 metres (66 feet).

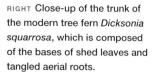

RIGHT Close-up of the trunk of the modern tree fern *Dicksonia squarrosa*, which is composed of the bases of shed leaves and tangled aerial roots.

BELOW Reconstruction of the Carboniferous tree fern *Psaronius* which strongly resembled modern tree ferns but grew to a height of over 20 metres (66 feet).

COAL SWAMPS

The club mosses, horsetails and ferns reached the peak of their success during the late Devonian and the Carboniferous periods, 380 to 290 million years ago. At this time the area of land that is now northern Europe and northeast America was actually a sinking freshwater basin situated near the equator. It had a warm, damp climate similar to that of modern tropical rainforests, which was ideal for plant growth. The huge club mosses, horsetails and tree ferns grew rapidly in the swamp conditions. *Lepidodendron* trees, for instance, took only around 10 years to reach their full height before dying and toppling back into the swamp. As well as the giant trees there was probably an understory of smaller trees and herbs, and the forest was also home to giant dragonflies, cockroaches and amphibians. As the trees died, they fell into the stagnant water, and plant debris accumulated faster than it could decay. Over a matter of centuries the material built up into layers of peat, before eventually being buried by other sediments and compressed to form coal.

Over a period of 100 million years the coal forests laid down the vast coal reserves that powered the Industrial Revolution. However, the swamps were so successful that they paved the way for their own destruction. Huge quantities of carbon became locked up in the coal, while roots exposed silicate rocks, which

BELOW The modern horsetails of the genus *Equisetum* reach a height of only 1.5 metres (5 feet). Fertile stems topped by reproductive cones emerge from the soil in spring.

absorbed yet more carbon dioxide. The result was that carbon dioxide levels plummeted. At the beginning of the Carboniferous period, the atmospheric carbon dioxide concentration was 20 times higher than it is today; by the end of the Carboniferous it was below today's levels. Carbon dioxide is a greenhouse gas, which keeps the Earth warm. Falling carbon dioxide levels caused the Earth to become cooler and drier, and caused the long Permo-Carboniferous ice age. The coal swamp giants could not survive in the changed climate and their modern descendants are a tiny remnant of a once great flora. Tree ferns do still exist, although only in isolated rainforest locations, but the other groups now contain only herbs. Today, the horsetails are represented by just a hundred or so species of the single genus *Equisetum*; they resemble drastically scaled-down versions of *Calamites*. The closest relatives of *Lepidodendron* are quillworts and the tiny club mosses.

EVOLUTION OF SEEDS

BELOW Reconstruction of the seed fern, *Medullosa*, showing its marked similarity to ferns in its habit and foliage.

Long before giant club mosses, horsetails and tree ferns had died out, other groups of trees had started to emerge. These had a suite of adaptations that allowed them to survive the drier conditions much better. The main difficulty the coal forest giants had in adapting to the drier conditions of the Permian was their method of reproduction. Like their modern relatives, they had two stages in their life cycle. The adult trees produced spores rather than seeds, which grew into tiny plants that lived on the thin film of water that coats the ground in damp areas. These tiny plants released sperm that had to swim to a neighbouring plant to fertilize the eggs the plant had produced. Only when this had taken place could a new tree develop, so damp conditions were crucial to completing the life cycle.

The new plants introduced an evolutionary novelty – the seed – that freed them from this reliance on surface water. Instead of producing one sort of tiny spore, they produced two different types: male and female. They held their large female spores on the parent plant, where they germinated and grew into seeds that were kept safe and moist within a protective coat. The new plants released only the smaller male spores, which were protected inside tiny cases or pollen grains. The pollen grains were transported by the wind to the female seed, germinated there, and produced sperm that only needed to swim a very short distance within the seed to fertilize the ovule within it. The fertilized seed could then be released. This method of reproduction is very reliable, particularly as the parent plant can provision the seed with a food store to help it to germinate and grow.

THE EVOLUTION OF SECONDARY THICKENING

A major evolutionary advance of the seed plants that further secured their success was the evolution of secondary thickening. The plants developed a layer of cells around the outside of their stems that could divide to produce tissue on both the inside and outside. This bifacial vascular cambium, as it is called, lays down woody tissue, or xylem, on the inside, strengthening the stem and increasing its water transport capability. On the outside it produces a thinner layer of phloem, a tissue that is used to transport sugars down to the roots.

This pattern of growth has two main advantages. First, it enables young trees to grow taller more rapidly. When they are small they can concentrate on extension growth, delaying thickening up their trunk until later in life, when it will be needed. Secondary thickening also makes branching easier, and these two factors make the young tree a far better competitor for light. Second, adding new wood allows trees to transport water up the trunk, even if the old xylem has been blocked during cold or dry weather. Trees with secondary thickening are therefore better adapted to seasonal climates than those without. In such climates the seasonal pattern is revealed by the growth rings in the wood. These rings occur because trees lay down a bigger proportion of wide, thin-walled cells in spring to promote water transport and narrower, thick-walled cells later in the year to promote mechanical stability. Growth rings can be seen in wood from as early as the Devonian period, showing that seasons existed then, as now.

The only problem with secondary thickening is that the delicate phloem tissue is exposed on the outside of

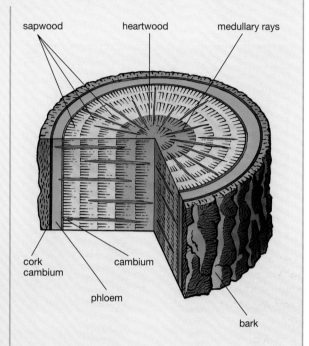

ABOVE **The trunk structure of a modern seed plant, which undergoes secondary thickening.**

the trunk and is vulnerable. Trees solve this problem by having a second cambial layer outside the main one. This cork cambium produces bark that protects the phloem from mechanical damage and herbivores, insulates the trunk from rapid temperature fluctuations and protects it from fire.

THE GYMNOSPERMS

The first seed plants were called the seed ferns because apart from having seeds they were, in fact, very similar to the ferns. The similarity is very apparent in *Medullosa*, a small tree common during the late Carboniferous period. However, the tree ferns soon gave rise to a very varied group of woody seed plants called the gymnosperms. Gymnosperms means 'naked seeds'; the group were so-called because the ovule is not fully enclosed. The gymnosperms dominated the land flora for the next 250 million years, just as reptiles dominated the land fauna. They therefore lived side by side with the dinosaurs – although, unlike the dinosaurs, many thrive to this day.

CYCADS AND CYCADEOIDS

The Cycads first appeared in the late Carboniferous period. They closely resemble tree ferns in their outward appearance, with a slow-growing cylindrical trunk that is supported and protected by old leaf bases, and a simple crown of palm-like leaves on top. Only a few cycads ever developed branched stems. The advances are all concentrated in their reproductive organs, which are large cone-like structures; these protect the seeds in female plants and the pollen grains in male plants. The pollen itself shows a second advance. When it lands on the female flower it produces a pollen tube that grows until it nearly reaches the female ovule. The sperm swim down the pollen tube and are released almost at their destination.

RIGHT A modern cycad, showing both the fern-like crown of leaves and the large reproductive cones that distinguishes it from ferns.

BELOW Reconstruction of the Jurassic cycadeoid, *Williamsonia*, showing the lack of branching and stem taper typical of these primitive gymnosperms.

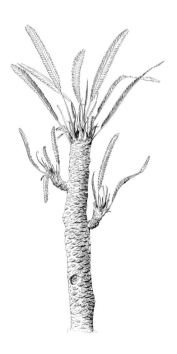

The cycads reached the height of their success in the warm Jurassic and Cretaceous periods, when they became the favourite food of such large herbivorous dinosaurs as *Triceratops*. Since then, however, they have declined, and only around 140 species survive today, mostly living in the tropics and subtropics. To this day, many are provided with formidable surface armour on their trunk, and tough, spiny leaves; these are perhaps a legacy of their attempts to ward off the giant reptiles!

A related group of seed plants, the cycadeoids, looked very similar to the cycads but produced one intriguing advance. They developed complex flower-like structures on their reproductive organs, which they must have used to attract insects. These plants may therefore have been the first to be insect-pollinated, and may have paved the way for the evolution of the true flowering plants. Despite having these flowers, though, cycadeoids have not survived. They declined after reaching a peak of diversity in the Jurassic and had died out, along with the dinosaurs, by the end of the Cretaceous period.

GINKGOS

The third group of gymnosperms, the ginkgos, were the first seed plants that were capable of extensive secondary thickening (see p. 17). They were therefore the first to resemble what we now think of as a 'typical' tree; they had a tall, tapering trunk, and a complex, branching crown. The ginkgos also had more modern-looking fan-shaped leaves. Like the cycads, they achieved their greatest diversity during the Jurassic period, when they had a worldwide distribution. They have since declined, and now just one species, *Ginkgo biloba*, remains. The ginkgo is limited in the wild to western China, but it is such an attractive tree that it has been planted in parks and in streets all around the world. The males, which have attractive catkin-like cones, are preferred to the females, whose fleshy seeds have a distinctly nasty smell.

ABOVE The characteristic fan-shaped leaves of the maidenhair tree, *Ginkgo biloba*, turn yellow and are shed in autumn.

CONIFERS

The most successful gymnosperms are the conifers. They first emerged during the Carboniferous period, became the favourite food of the giant sauropod and duckbilled dinosaurs during Mesozoic times, and continue to thrive to this day. There are around 570 species and they cover large areas of the temperate and subpolar regions. The group made key advances in their reproduction, further freeing them from the need for water. They do not rely at all on swimming sperm; instead the pollen tube discharges the male nuclei directly into the ovule. The reproductive structures of conifers are also well protected, in most cases by the cones that give the group its name.

The other characteristic for which conifers are well known – needle-like foliage – is not found in all conifers, but is a relatively recent innovation. The first conifers actually had leaves that looked more like those of modern broad-leaved angiosperm trees. The same is true of members of the primitive family Araucariaceae, which first appeared in the Triassic period and thrived in the early Jurassic. They have left numerous remains, including jet found on the coast of Yorkshire in the UK, which is in fact fossilized wood. Other fossils show that these trees were remarkably similar to their few living descendants. These modern-day Araucariaceae include the so-called 'living fossils', such as the monkey puzzles of South America and Australasia and the kauri pines of Australasia and Southeast Asia. Both these types of tree have very tall, straight stems, and regular whorls of branches; these characteristics give them a strange, primeval look, accentuated in the monkey puzzle by its regularly arranged scale-like leaves. Unfortunately, the straightness of the stems makes these trees a very valuable timber crop, so most of the species in this group are threatened by logging operations. The podocarps, of the family Podocarpaceae, are another ancient group that often have broad leaves. Like the Araucariaceae, they are also mostly restricted to the southern hemisphere.

BELOW The yew, *Taxus baccata*, has typical drought-resistant needles. However, unusually for a conifer, its seeds have a fleshy outer coat, attracting birds to disperse them.

A third family, the redwoods of the family Taxodiaceae, probably emerged later, at the beginning of the Jurassic period. The fossil record shows that they used to have a worldwide distribution; for most of the last 60 million years, swamps dominated by redwoods thrived in the temperate regions of the world, which had a warmer climate than they do today. It is these swamps that formed the brown coal deposits of central and eastern Europe. However, by the end of the Tertiary, two million years ago, the redwood forests had all but disappeared; the cooling of the climate and the start of the ice ages had limited the swamps to a few isolated areas with more equable climates. Today the redwoods are represented by just a few species: the dawn redwood of China, the swamp cypresses of the southeast USA, and the two giant redwoods of the southwest USA. The yews of the family Taxaceae, with their fleshy fruit-like seeds, and the true cypresses, with their scale-like foliage, are more widespread. Perhaps these characteristics better suit them to the drier world of today.

The most successful group of conifers is the family Pinaceae, which includes around 300 species of pines, larches, spruces, firs and cedars. The members of this group have narrow needle-like leaves that are ideally suited to resist drought, fire and frosts. They have therefore long dominated the colder parts of the northern temperate zones. This family has probably benefited from the recent (in geological terms!) cooling and drying of the world's climate; it has helped them spread even further over the northern hemisphere. Boreal forests of fir, pine, spruce and larch cover huge areas of the subarctic, while evergreen forests of pines and cedars cover much of the drier temperate areas, including the Mediterranean. The larches are among the few deciduous conifers, and they manage to inhabit many of the cooler and drier areas of Canada and Siberia.

THE ANGIOSPERMS

The flowering plants, or angiosperms, are the most recent of all the major plant groups. They are also by far the most successful; there are some 275,000 species. However, in contrast to the gymnosperms, the majority of species are not trees but herbs. Even so, there are still around 80,000 species of angiosperm trees (compared with only 600 species in the other groups), making them by far the most diverse group. The first angiosperms are thought to have been small trees or shrubs that evolved at the beginning of the Cretaceous period in warm tropical forests. As we shall see, the advances they made helped them not only to dominate this sort of habitat but also to invade new habitats and assume a whole range of body shapes.

ADVANCES OF THE ANGIOSPERMS

The angiosperms made advances over the gymnosperms in both their vegetative structure and their reproduction. The most important structural alteration was in their wood. In gymnosperm wood there is only one sort of xylem cell, which has two functions: to transport water up to the leaves and to strengthen the trunk. This is a problem because water transport is improved by having wide cells, whereas strength is improved by having long, narrow cells. The cell shape has to be a compromise between the two. Gymnosperms have very long, narrow cells called tracheids, which limit their water-conducting ability. The angiosperms developed wood that is differentiated into two main types of xylem cells: long, narrow tracheids, or fibres, which strengthen the stem and wide, thin-walled vessels, which pipe water through the trunk. Since the resistance of pipes to flow falls dramatically as they get wider, angiosperms therefore have greatly improved water transport, even though the vessels make up only a small proportion of the cells.

Angiosperms improved their reproduction in three major ways. First, they fully enclosed their seeds (hence the name angiosperm, which means 'container seed'); this protects the seed from drought. Second, they speeded up their seed production by having a smaller ovule that only laid down its energy reserves after fertilization. Third, they improved their pollination efficiency, compared with the wind-pollinated gymnosperms, by developing flowers that attracted animal pollinators. The flowers produced not only pollen that the insects could eat, but also nectar, and they advertised its availability with showy petals and enticing scent.

BELOW LEFT Scanning electron micrograph (x100) of the wood of the angiosperm *Quercus robur*, showing a large diameter vessel (top) within the narrower tracheids. The grain runs vertically, parallel to the long axis of the trunk.

BELOW Scanning electron micrograph (x100) of the wood of the gymnosperm *Agathis australis* showing the uniformly narrow tracheids. The grain runs vertically, parallel to the long axis of the trunk. Sections through some rays can also be seen.

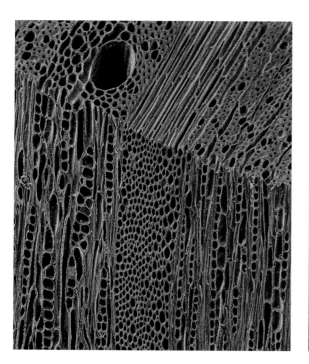
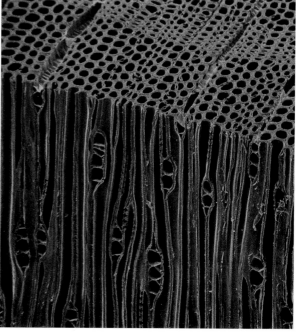

EVOLUTIONARY RADIATION OF THE ANGIOSPERMS

The first flowering plants probably looked rather like modern magnolias; they were small trees with thick evergreen leaves and large open cup-shaped flowers pollinated by beetles. These trees soon started to take over from the gymnosperms, and also evolved into plants that had a range of body plans and life histories. Conifers could not develop into climbers, because they would simply not be able to transport water fast enough up through narrow stems to feed their leaves. The evolution of vessels allowed angiosperms to develop thin but hydraulically efficient stems, enabling them to enter this niche. Today, tropical rainforests are full of woody climbers called lianas, and around 10 per cent of all angiosperms are climbers. Vessels may have also helped angiosperms evolve into herbs; they would allow a turgor-supported stem to transport adequate water up to the leaves even if the stem only contained a few strands of xylem. A further factor aiding the evolution of herbs was that angiosperms could rapidly produce seeds and therefore could reproduce even before they had started to become woody. The short-lived herbs that began to emerge soon started competing with, and taking over from, herbaceous ferns beneath the tree canopy. Today, herbs make up the great majority of angiosperm species.

Because the generation time of herbs is so much shorter than that of trees, they can evolve much faster. Therefore, the majority of evolutionary novelties are in the flowering plants, as their emergence probably took place in herbs. Nowadays herbaceous angiosperms have successfully invaded deserts, marshes, fresh water and even the open sea, and they have evolved a veritable cornucopia of defensive chemicals.

However, despite the success of herbs and climbers, being a tree remains one of the best ways to compete for light; many of the families of flowering plants therefore include members that have evolved from herbs back again into trees. The rowans (mountain-ash) and cherries of the rose family, and the locust trees and laburnums of the pea family, are just a few examples of this trend.

BELOW The primitive radially symmetrical flower of a Magnolia, showing the petals surrounding the spirally arranged male stamens, and the female carpels at the centre of the flower.

MONOCOT TREES

So great are the advantages of being a tree that trees have even evolved in a group of flowering plants, the monocotyledons (or monocots), that long ago lost the ability to undergo secondary thickening. It is thought that the first monocots may have been aquatic plants that distributed their woody tissue in isolated strands throughout their body to better resist stretching by water currents. Unfortunately, though, this also prevented a vascular cambium forming, so few

LEFT The Australian 'grass tree' *Xanthorrhoea australis*. The bases of the grass-like leaves it produces at the tip eventually get packed together to produce the trunk.

BELOW The base of the trunk of a screw pine, *Pandanus sp.*, showing the prop roots that emerge from and support the thin basal regions of the trunk.

monocots can grow thicker by laying down wood as they get older. The monocots include many narrow-leaved herbs such as grasses, lilies and orchids; however, many trees have evolved in this group, all of which produce their trunks in unusual ways.

The palms produce their trunks by using not secondary thickening but primary thickening; they simply make the tip of their stem very thick and fill it with woody tissue. The disadvantage is that the base of the trunk has to be as wide when it is laid down as it will be when the plant reaches its maximum height. This makes palms very slow growing at first compared with normal trees. They also lack the ability to branch properly because they have just one terminal bud; consequently, they have only a single crown of leaves, just like tree ferns. Nevertheless, they have proved very successful in tropical and subtropical areas, and today there are over 3,000 species of palms. A similar growth pattern is used by the 'grass trees' of Australia, which resemble pillars with grasses growing on top.

The screw pines or pandanuses of the Old World tropics use a different method to support themselves as they grow. They increase the diameter of the tip of the stem as they get taller, and support the thin lower regions of their trunk by developing long, thick aerial roots that prop up the plant. The dragon trees of North Africa and the Canary Islands and the Joshua trees of North America have even managed to evolve a form of secondary thickening rather like that of the giant club mosses of Carboniferous times; they lay down a bark-like material in their outer cortex as they grow. This allows them to produce trees that have more mechanically efficient tapered trunks and large branching crowns.

BELOW The Joshua tree, *Yucca brevifolia*, showing the strap-like evergreen leaves that help it survive in the desert by reducing water loss.

Two other types of monocotoyledon trees have also evolved into tall plants, although they are not generally regarded as trees. First, the bamboos which are grasses with hollow, woody stems that can grow to heights of up to 40 metres (131 feet). Unlike true trees though, a single plant of bamboo produces many such stems that lean against each other for additional support. Secondly, banana plants also look like trees, but these are in fact huge herbs, the stems of which emerge from a thick underground rhizome. Although banana plants are tall, their stems are very narrow, and these are supported by the fleshy bases of leaf stalks surrounding them.

LEFT The thick stems in a stand of the giant bamboo, *Phyllostachys nigra*, lean against each other for added support.

CHAPTER 2

How trees lift water

IN MANY REGARDS TREES WORK in just the same ways as other land plants. The chloroplasts in their leaves trap light energy from the Sun by photosynthesis, and they use this energy to convert carbon dioxide and water into sugars and oxygen. Just as in other plants, the carbon dioxide enters the hollows of the leaf during the day, through open pores, called stomata. The open stomata also let water vapour escape, in a process called transpiration. This water, together with the much smaller amount actually needed for photosynthesis, is replenished by water from the soil; the water is taken up by the root system and piped up the xylem tissue to the leaf, through the roots, stem and branches. Sugars are in return transported down the plant along the phloem tissue. Much of the sugar is stored in the ray tissues of the branches, stem and roots, and the sugars are also used to power growth in these structures, which cannot photosynthesize.

The differences between trees and other plants are really ones of scale. In particular, two main questions have troubled botanists. How can 100-metre (328-foot)-high trees lift water all the way to the top of their canopy? And how can such huge structures stand up against their own massive weight, and the huge forces imposed on them by the wind? This chapter and the next one describe how biologists have answered these questions, concentrating on the commonest and best-studied groups of trees, the conifers and the angiosperms.

OPPOSITE An emergent *Koompassia excelsa* rising above the rainforest canopy in Sabah, Malaysia. Because of their great height, emergent trees can support only a sparse canopy of leaves, allowing light to penetrate to the main canopy below.

BELOW The physiology of a typical tree, showing photosynthesis in the leaves and the movements of water, nutrients and sugars to and from them.

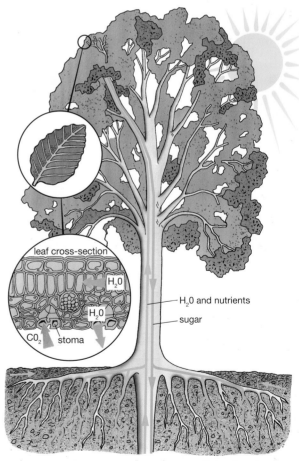

leaf cross-section

H_2O

H_2O

CO_2 stoma

H_2O and nutrients

sugar

H_2O and nutrients

WOOD STRUCTURE AND WATER TRANSPORT

Even a quick look at the magnified sections of pieces of wood in the images on p. 23 and below suggests that its structure is well suited to transport water up from the roots to the leaves of a tree. Over 90 per cent of the wood cells are arranged along the axis of the trunk or branch, like thousands of closely packed drinking straws. Water can flow through them up the tree. It is much harder to work out what drives the water upwards. Over the last two centuries several possible mechanisms have been suggested, but only one has stood up to experimental investigation.

The first suggestion is that water is actually pumped up the trunk by the roots using some sort of osmotic mechanism. Unfortunately, this idea does not live up to close scrutiny. If water were pumped up from below one would expect water to seep, or even spray, out of a tree trunk when it was cut down. This hardly ever happens, so this cannot be the normal mechanism. Positive 'root pressure' *is* produced by birches and maples, but only in early spring. As we shall see later, these trees do this for a specific reason.

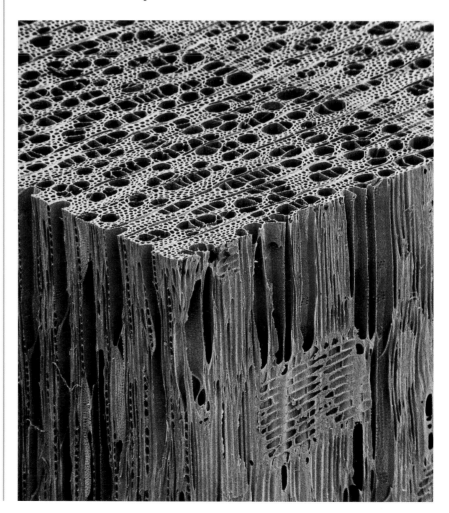

RIGHT Scanning electron micrograph (x60) of the wood of the beech, *Fagus sylvatica*, showing the 'drinking straw' vessels packed between the narrower fibres, or tracheids, and oriented parallel to the long axis of the trunk.

A second suggestion is that water could be dragged up a tree by capillary action, just as soil in a plant pot draws up water from a saucer underneath. The problem with this idea is that capillary action only works with very thin pipes. Wood cells are simply not narrow enough to draw water very far against gravity; with their diameters of between 30 and 300 micrometres (1.2 and 12 thousandths of an inch) they would only be able to draw it up to height of 5–50 centimetres (2–20 inches). A similar problem means we must reject a third idea: that water is dragged up by sucking it from above, just as we draw water up a drinking straw. When we suck on a straw we lower the pressure at the top of the straw, and the water is forced up by atmospheric pressure, which pushes on the other end of the straw. This technique cannot raise water very far either. Even if trees could produce a vacuum in their leaves, atmospheric pressure would only apply a pressure of one atmosphere (14 pounds per square inch, or p.s.i.), enough to raise the water to a height of just 10 metres (33 feet).

THE COHESION THEORY

The surprising solution to the problem is that water is actually lifted up from above; it is pulled up under tension as water is lost from the leaves by transpiration. When this was first suggested in 1894 the idea was greeted by disbelief, but since then a large amount of evidence has been found to support it. For a start it has been shown that if water is held in a narrow pipe it can actually withstand large stretching forces without breaking, just like an elastic band. The water's strength is due to the cohesion between its molecules and can be readily demonstrated by experiment. If capillary tubes are filled with water they can be spun around at high speeds in a centrifuge; the water will resist the centrifugal force, which would tend to make the water spray out of the ends of the tube. It can withstand stretching forces of up to 280 atmospheres (more than 4,000 p.s.i.). This is 10 times the strength of a typical elastic band. The cohesive strength of water could therefore hold up a column of fluid nearly three kilometres (1.9 miles) high.

It is also possible to show that water in a tree is being stretched. When wood vessels are cut you can actually hear the hissing sound of air entering. You can measure the stretching force by performing a second experiment. If a branch is cut from a tree, the water column is broken and the water shortens, just as an elastic band does if it breaks. The water column retreats into the branch. The branch can then be sealed with its leaves held in a pressure vessel (called a 'Scholander bomb' after an early tree physiologist) and the cut end sticking out. A pressure can then be applied within the vessel, which squeezes the shoot, forcing the water out of the cut end again. Water starts to come out of the cut end when the positive pressure applied equals the tension that the water was under. Using this technique, stretching forces of well over 20 atmospheres (294 p.s.i.) have been detected;

this is twice as much as is needed to raise water to the top of even the highest tree. The remaining pressure overcomes the resistance of the narrow tracheids and vessels to water flow. The water is under so much tension, in fact, that it can measurably deform the tree; strain gauges mounted on the trunk have shown that during the day (when trees are transpiring) a 20-metre (66-foot)-high tree with a trunk diameter of 30 centimetres (12 inches) will shorten by about one centimetre (0.4 inches) and get one millimetre (0.04 inches) thinner.

THE HYDRODYNAMIC DESIGN OF WOOD

The cohesion mechanism seems to be very effective in raising water up trees. There are only two disadvantages with this method. First, because trees need to lose water from their leaves to keep lifting it up, they take up far more water than they actually use for photosynthesis; a large tree can take up more than 500 litres (110 gallons) of water a day. This means that trees tend to dry out the soil beneath their canopies, and makes them more vulnerable to droughts. On the positive side though, evaporation of water cools the leaves down, and prevents them overheating in the sun.

A second problem is that water under tension is very unstable, and air bubbles within the wood can spell disaster. Small bubbles are not a problem; they will collapse as a result of the surface tension in the water and be reabsorbed. However, if large bubbles get into a wood vessel the water column will be broken by the tension in the water above and below it. The whole vessel will then fill with air, forming what is called an embolism. Once an embolism has formed it can be prevented from spreading the whole length of the tree by incorporating sieve-like plates along the wood vessels; these plates trap the air bubbles. However, unless water is actively forced into the vessel again, the vessel will remain empty and will have lost its conducting ability. You can detect the formation of embolisms in an actively transpiring tree by attaching a microphone to its trunk; the breaks of the water columns make little clicks. Embolisms are common in very dry conditions, when there is little water to drag up from the roots, but the greatest danger is from frost. As water freezes, dissolved air comes out of solution, and air bubbles form. You can see these sorts of bubbles in the ice cubes in your freezer. When the ice thaws these bubbles can expand and rapidly empty a vessel.

The problem with embolisms means that the design of wood has to be a compromise. To reduce the resistance of the wood to water flow it is best to have wide, open-ended conducting cells; in contrast, to minimize the chances of vessels emptying because of embolisms it is best to have short, narrow conducting cells. The two main groups of trees, the conifers and the angiosperms, operate at different ends of the spectrum in this compromise between efficiency and safety.

WOOD OF CONIFERS

The wood of conifers seems to be adapted to be safe rather than efficient at transporting water. It is made up of many thin tracheids, each of which is about 30 micrometres (1.2 thousandths of an inch) wide and between 0.1 and 10 millimetres (0.004 and 0.4 inches) long. Each tracheid has a closed, tapered end and is joined to its neighbours only by a number of tiny holes called bordered pits. The consequence of this design is that conifers rarely suffer from embolisms, and the embolisms that do form rarely spread between tracheids. Conifer wood is therefore ideal for trees that grow in areas with long, cold winters or dry, hot summers; even in spruce trees growing in the subarctic, the wood only loses two per cent of its conducting ability each year. This loss can easily be replaced by adding new wood to the trunk as the tree grows.

LEFT A transverse section of the wood of Sitka spruce, *Picea sitchensis*, (x30) showing the masses of narrow tracheid cells. There are also a few scattered pore canals along which resins are transported. The section shows just over two growth rings.

BELOW Bordered pits (x500) in the side walls of the tracheids of *Pinus radiata*. Water has to travel through many such pores on its course up the tree.

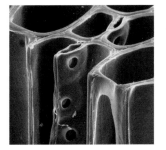

WOOD OF ANGIOSPERMS

The wood of angiosperms seems to be designed to be efficient at conducting water rather than safe. It has lots of wide vessel cells that can be up to 300 micrometres (12 thousandths of an inch) in diameter. These cells abut end to end with other cells to produce pipe-like vessels that can be several metres long. The consequence is that angiosperm wood transports water very efficiently, but is very prone to embolism in dry or freezing weather.

The design of angiosperm wood is ideal for life in tropical rainforests, where the warm, wet climate reduces the risks of embolism. The high conducting ability means that even a narrow-trunked tree can conduct enough water to supply its leaves. However, angiosperm trees are also common in subtropical areas that are prone to seasonal drought, and temperate areas that have cold winters. In these regions the vessels will inevitably suffer catastrophically from embolism; so how do the trees survive? The answer is that angiosperms use two very different methods of coping.

Trees like oaks and ashes put up with the loss of all their vessels each winter, relying only on the current year's wood to transport the water. The wood they produce in spring contains many large vessels, some 300 micrometres (12 thousandths of an inch) in diameter, and they use these vessels to supply their new leaves with water. Later in the year, the wood they produce is mostly composed of mechanical fibres and contains only a few narrow vessels. Their wood is described as 'ring porous', because it has rings of more porous tissue between rings of dense mechanical tissue. The strategy is successful but has two disadvantages. First, the trees cannot break bud until after the first new wood is formed. This is why oaks and ashes are among the last trees to burst into leaf each spring. Second, these trees are also vulnerable to late frosts that can embolize their new vessels. Indeed, in northern Europe and the northeast USA these trees are at the northern extremes of their range, and most of their relatives live in the tropics.

Other trees, such as poplars, beeches, birches and maples, have 'diffuse porous' wood. This contains large numbers of rather narrower vessels, with diameters of 60 to 100 micrometres (2.4 to 4 thousandths of an inch), which are spread more or less evenly throughout the wood. These narrow vessels are less prone to embolisms than the wide vessels of oaks. The trees also prevent embolisms spreading by separating adjacent vessel cells with rows of bars called 'scalariform plates'; these trap the air bubbles. Nevertheless, there is still a relatively high rate of embolism. Poplars seem to just put up with it; birches and maples, in contrast, use a clever strategy to reverse embolism. They refill their empty vessels each spring by using their roots to pump sugar-rich water up into their trunks and branches. This strategy helps birches and maples survive extremely well in cold northern climates, and some birches can even live in the arctic tundra. The strategy has been exploited commercially; both maples and birches can be tapped in spring to collect the sugar-rich liquid, which is boiled down to produce maple syrup and birch sap wine.

BELOW A transverse section of the wood of ash, *Fraxinus excelsior*, (x30). In this ring porous angiosperm the wide vessels are located towards the inside of each growth ring, and are surrounded by narrow fibres.

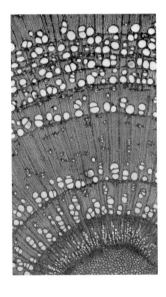

RIGHT A transverse section of the wood of beech, *Fagus sylvatica*, (x30). In this diffuse porous angiosperm the narrow vessels are dispersed throughout the growth ring. The dark radial bands are large rays.

MYCORRHIZAS AND NUTRIENT CAPTURE IN TREES

Trees obtain their nutrients from their roots just like other land plants do. However, their roots are so thick that they are very inefficient at mining through the soil for the scarce nutrients. In order to obtain sufficient nutrients, they would need to produce large amounts of root tissue. Trees get over this difficulty by having a symbiotic relationship with soil-dwelling fungi. The fungi grow in a sheath around the roots, their narrow hyphae ramifying through the soil and absorbing large amounts of nitrate and phosphate. The fungus passes some of these on to the tree roots, receiving carbohydrates in return. Almost all trees have these so-called ectomycorrhizas (meaning 'outer root fungi'), and in fact a great number of the mushrooms that emerge in woodland are the fruiting bodies of such organisms that are not harmful, but positively beneficial, to the tree.

SAPWOOD AND HEARTWOOD

Whatever the type of wood, it inevitably loses some of its ability to conduct water as it ages. Trees therefore progressively 'shut down' old wood, to produce a dead core of darker material in the centre of the trunk. This is called heartwood, in contrast to the water-conducting sapwood, which is lighter in colour and contains living cells. As heartwood is formed the vessels are blocked and the wood cells are filled with resins and gums. These repel disease fungi and wood-boring insects and stiffen up the cell wall. All the remaining wood cells in the developing heartwood then die, but the cell walls are preserved and help support the tree for many years to come.

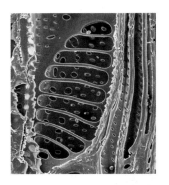

ABOVE Scalariform plate (x300) between two vessel elements in the New Zealand privet, *Griselinia littoralis*. This prevents embolisms spreading along the vessel.

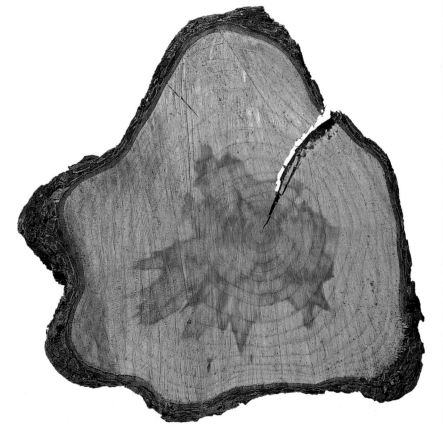

LEFT Section through the base of the trunk of a larch, *Larix europea x L. japonica*, showing the darker heartwood in the centre.

CHAPTER 3

How trees stand up

THERE IS NO DOUBT THAT TREES are magnificent structures; a mature coastal redwood tree would overshadow most church steeples. Like all engineering structures, trees combine two elements to do this: they use good materials and they arrange the materials so that they are used to their best advantage. Trees have only one structural material – wood – but as we shall see this is superbly engineered. Trees are also ingeniously designed structures that combine strength and flexibility. They can even respond to their environment and change their design accordingly. This allows them to support their canopy of leaves using the bare minimum of wood.

THE MECHANICAL DESIGN OF WOOD

Wood needs to combine many useful properties to allow it to support the leaves of trees. It has to be stiff, so that trees do not droop under their own weight; it has to be strong, so that the sheer force of the wind does not snap the trunk and branches; it has to be tough, so that when the tree gets damaged it does not shatter; and finally it has to be light, so that it does not buckle under its own weight. No manufactured material could

OPPOSITE **The main function of buttresses is to help in anchorage. In this tree from Queensland, Australia, the buttresses join the trunk like angle brackets to the lateral sinker roots.**

BELOW **Forces that are set up and movements that occur when a tree is blown sideways by the wind.**

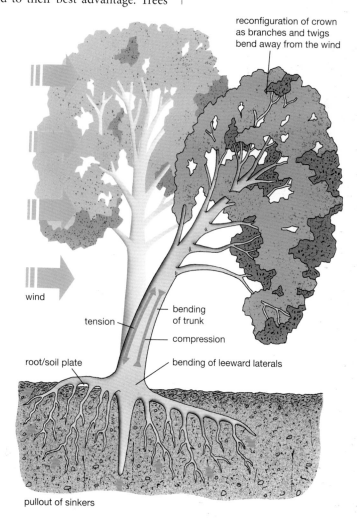

reconfiguration of crown as branches and twigs bend away from the wind

wind

tension

root/soil plate

bending of trunk

compression

bending of leeward laterals

pullout of sinkers

do all of these things: plastics are not stiff enough; bricks are too weak; glass is too brittle; steel is too heavy. Weight for weight, wood has probably the best engineering properties of any material, so it is not surprising that we still use more wood than any other material to make our own structures! Its superb properties result from the arrangement of the cells and the microscopic structure of the cell walls.

ARRANGEMENT OF CELLS

Over 90 per cent of the cells in wood are long, thin tubes that are closely packed together, pointing along the branches and trunk. As we saw in the last chapter, this helps them transport water to the leaves, but it is also ideal for providing support. This is because they point in the direction in which the wood is stressed.

Trees mainly have to resist bending forces. Their branches have to resist being bent down under their own weight, and both the trunk and branches have to resist being bent sideways by the wind. These bending forces actually subject the wood inside to forces that are parallel to the trunk or branch; the concave side is compressed, while the convex side is stretched. Whichever way the tree is bent, therefore, the internal forces always act parallel to the cells or 'grain' of the wood. The long, thin wood cells are well suited to resist the forces; the cells on the concave side resist being compressed, rather like pillars, while those on the convex side resist being stretched, rather like ropes. As a consequence, wood is very strong along the grain.

The cellular nature of wood is also advantageous to the tree for another reason. Because the cells are hollow, the tree's trunk and branches can be thicker than if all its wood material was laid down in a solid mass. (In some trees, such as the tropical pioneer *Cecropia*, not only the cells but also the trunk and branches are hollow.) Weight for weight, hollow structures like these are stronger than solid structures; this is why tubes are so often used in large engineering structures.

ABOVE *Cecropia* trees, like this one in the rainforests of Belize can have such a slender trunk because of the efficiency of its lightweight design (below).

RIGHT The split trunk of *Cecropia* showing its lightweight tubular structure and strengthening nodes or bulkheads.

RAYS

The arrangement of the cells along the trunk does have one potential disadvantage. It is relatively easy to split wood parallel to the trunk, what a carpenter would call 'along the grain'. However, this is not very important to the tree because its wood is hardly ever subjected to forces in this transverse direction. As an extra precaution, trees prevent the wood splitting between successive growth rings by incorporating into it blocks of cells called rays, which are oriented radially in the trunk. As well as storing sugars, these rays act rather like bolts, effectively pinning the wood together. However, the rays do not prevent the wood from splitting radially from the centre of the trunk. This is also why the easiest way to cut up wood with an axe is radially, through the centre of the trunk, like cutting pieces of pie. It is also the reason why it is near impossible to break off a living branch of a tree; the branch breaks halfway across easily enough, but then splits along its length through the centre, a pattern of fracture known as 'greenstick fracture'.

secondary wall

inner layer (S$_3$)

middle layer (S$_2$)

outer layer (S$_1$)

primary wall

middle lamella

ABOVE **The cell wall structure of wood fibres or tracheids. The bulk of the cellulose microfibrils are in the S$_2$ layer, where they are arranged at 20 degrees to the long axis of the cell.**

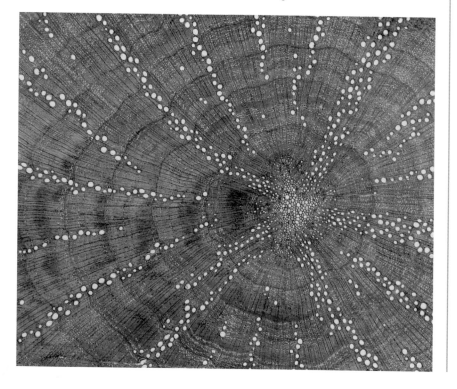

LEFT **Medullary rays in a *Quercus ilex* branch. They are the dark lines of cells radiating from the centre.**

STRUCTURE OF THE CELL WALLS

The structure of the cell walls also improves the mechanical properties of wood. Cell walls, like fibreglass, are a composite material. They are made of tiny cellulose microfibrils, which are embedded in a matrix of hemicellulose and lignin. The cellulose fibres stiffen the material, like the glass fibres in fibreglass, while the matrix protects the fibres and prevents them from buckling, like the resin in fibreglass. This gives the composite a combination of high stiffness and strength.

Embedding fibres within a matrix also improves the toughness of composite materials because more energy is needed to break them; the energy is used up pulling the fibres out of the matrix. (For this reason fibreglass is around a thousand times tougher than either resin or fibres on their own.) The arrangement of the fibres within the walls of wood cells helps to make wood even tougher. Wood cells have walls with several layers, but the thickest layer, making up 80 per cent of the wall, is the so-called S_2 layer. Here the microfibrils are arranged at an angle of around 20 degrees to the long axis of the cell, winding round the cell in a narrow helix. This is not far off being parallel to the cell wall, so they stiffen it up along the grain quite effectively. But the greatest effect is to dramatically increase the toughness. As the wood is stretched, the cells do not break straight across; instead, the cell walls buckle parallel to the fibres and the different strips of the cell wall are then unwound like springs. This process creates very rough fracture surfaces and absorbs huge amounts of energy, making wood around a hundred times tougher even than fibreglass. This mechanism only acts when wood is cut across the grain, but it explains why wooden boats are far sturdier than fibreglass ones and can absorb the energy in minor bumps without being damaged.

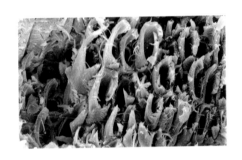

ABOVE Fracture structure in Sitka spruce wood. Note the rough helical fracture of the cell walls, and the loose fibre strands.

PRE-STRESSING OF WOOD

Wood has just one problem: because wood cells are long, thin-walled tubes, they are very prone to buckling, just like drinking straws. This means that wood is only about half as strong when compressed as when stretched, as the cells tend to fail along a so-called compression crease. If you bend a wooden rod the compression crease will form on the concave side and subsequently greatly weakens the rod. Trees prevent this happening to their trunks and branches by pre-stressing them.

New wood cells are laid down on the outside of the trunk in a fully hydrated state. As they mature their cell walls dry out and this tends to make them shorten. However, because they are already attached to the wood inside, they cannot shrink and will be held in tension. Because this happens to each new layer of cells, the result is that the outer part of the trunk is held in tension, while the inside of the trunk is held in compression. The advantage of this is that when the trunk is bent over by the wind, the wood cells on the concave surface are not actually compressed but some of the pretension is released. It is true that on the other

LEFT This plane (sycamore) tree, *Platanus occidentalis*, has suffered from brittleheart. As a result its trunk has become hollow and it is vulnerable to breaking in high winds

convex side the cells will be subjected to even greater tensile forces, but they can cope very easily with those. The consequence is that tree trunks can bend a long way without breaking. This fact was exploited for centuries by shipwrights, who made their masts as far as possible from complete tree trunks.

ABOVE The trunk of this oak split when it was cut down due to excessive pre-stressing, reducing its value as timber.

Pre-stressing has two unfortunate consequences. Many trees are prone to a condition known as 'brittleheart'. This occurs because as the wood in the centre of the tree ages it can be attacked and broken down by fungi. Eventually it becomes so weak that the precompression force makes it crumble, and the tree trunk becomes hollow. Another problem occurs when trees are harvested. Cutting the trunk frees the cut end, and in some species this allows the pre-stress to be relieved; the centre of the trunk extends and the outside contracts, bending the two halves of the trunk outwards and causing the trunk to split along its length. These splits are known to foresters as 'shakes' and render the timber useless. In some fast-growing species of *Eucalyptus* the trunk can spring out so violently that it can kill the lumberjack who is cutting it down.

THE MECHANICAL DESIGN OF THE SHOOT SYSTEM

As we saw in the first chapter, there are essentially two parts to the shoot systems of trees: a rigid trunk and a flexible crown of branches, twigs and leaves. This combination of rigidity and flexibility plays a key part in helping trees stand up. In actual fact, it is usually the wind that is most likely to destroy a tree, or in some areas the weight of snow. Trees do not collapse under their own weight, unlike some of the structures made by humans!

WITHSTANDING THE WIND
Trees use a single trunk rather than many separate stems for the same reason that we use single poles to hold up flags; weight for weight, one thick rod is better at resisting bending than several thin ones. As a result, a single trunk can support a crown of leaves using a minimum of wood. Like flagpoles, tree trunks are also tapered; they are thickest at their base where the bending forces are greatest, but progressively thinner towards the tip. This also helps to minimize the amount of wood they use.

RECONFIGURING IN THE WIND
The trunks of mature trees are too rigid to bend far away from the wind. Fortunately, because the branches and twigs are so much thinner, the whole crown of the tree can. This bending of the crown makes it much more streamlined, reducing the aerodynamic drag force that it transmits to the trunk. Wind-tunnel tests have shown that this process of 'reconfiguration' can reduce the force on a five-metre (16-foot)-high pine tree in high winds to under a third of what it would be if the tree were rigid. Angiosperm trees can perform even better than conifers in this respect. Palm trees can bend right over in the wind and so withstand even the strongest hurricanes.

Wind-tunnel tests on deciduous angiosperms have shown that their leaves can reconfigure as well as their branches; they roll up in the wind to form streamlined tubes, which greatly reduces their drag. The leaves that do this best are lobed leaves, such as those of maples, and pinnate leaves, such as those of ash. However, even in trees such as oaks or hollies that have stiff leaves, the drag is reduced because the rigid leaves are flattened against the branches. Unfortunately, no research has been done to determine just how efficient the reconfiguration of full-sized angiosperm trees is at reducing their drag. They just don't fit into wind tunnels!

THE MECHANICAL DESIGN OF BARK

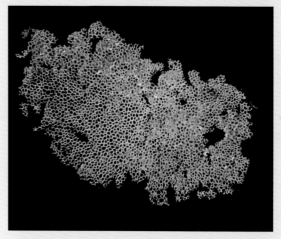

LEFT **The bark of the cork oak, *Quercus suber*, is particularly thick and is harvested commercially.**

ABOVE **A transverse section through cork (x30) showing that it is composed of numerous dead, air-filled cells.**

Bark acts as a superb shock-absorber, protecting the delicate phloem tissue from damage. The key to this ability is that bark is mostly composed of cork, which has a most ingenious structural design. Cork is made up of large numbers of closely packed cells, each of which is dead and filled with air. Each cell is a hexagonal prism in shape with side walls that are corrugated, like the walls of an open concertina. Because of the corrugation, a small crushing force can readily cause the cells to flatten out, like a closing concertina. Each cell can collapse to only a quarter of its original thickness, so this process can absorb a great deal of energy, safely dissipating impacts. This is good for the tree but even better for us. The properties have proved to be ideal to produce a stopper that is watertight yet easy to insert and remove. Real corks are still better in this respect than the artificial corks that have been recently introduced by winegrowers. Cork is produced sustainably by harvesting the thick bark of the cork oak, *Quercus suber*, which grows on the Iberian peninsula. The cork is cut from the tree every 10 years or so, without apparently damaging the living trees; they recover and just produce more cork. Cork has also been used to make flooring, where its shock-absorbing characteristics stop it becoming slippery.

SHEDDING SNOW

The conifers that grow at high latitudes or high altitudes have a crown design that allows them to shed snow. They are conical in shape, and both the main branches and side branches of firs point downwards before curving gently upwards like a ski jump ramp. Snow simply slides off these branches before its weight can damage the tree.

BRANCHES AND FORKS

The joints of trees, their branches and forks, are potentially vulnerable areas, just like joints in man-made structures. Branches are usually fairly strong, as the base of the branch is enclosed, in the form of a knot, within the trunk, but forks could potentially split apart very easily. Recent research we have carried out has shown that the wood at the apex of the join between the two branches is altered to provide reinforcement. It has a high density, and the wood fibres going up the two branches wrap around each other to hold the joint together. Early people clearly knew about this, and the handles of bronze age axes, the framework of their wooden ploughs and the keels of ships are often constructed from the branch junctions and forks of trees.

THE MECHANICAL DESIGN OF THE ROOT SYSTEM

Despite the reconfiguration of their crowns, trees still transmit large wind forces to their trunks and down to their root system. Fortunately, the root systems of most trees are well designed to anchor them firmly in the soil. The root systems of young trees are dominated by their tap roots. These anchor the trees directly, like the point of a stake. The rest of the anchorage is provided by the lateral roots, which radiate sideways out from the top of the tap root; they act like the guy ropes of a tent, stopping the tap root from rotating.

As trees get older, the tap root becomes less important. Instead, the lateral roots, many of which grow straight out of the trunk, start to dominate the root system; they get much longer and thicker, branching as they grow. They produce a network of superficial roots that ramify through the topsoil, sometimes two or three times further out than the edge of the crown. The lateral roots are well placed to absorb nutrients from the humus layer at the soil surface, but not to take up water in times of drought; neither are they well orientated to anchor the tree. Trees overcome these deficiencies by developing sinker roots that grow vertically downwards from the laterals, usually quite close to the trunk. If a tree is pushed over, a plate of roots and soil is levered upwards, about a hinge on the leeward side of the trunk. Some anchorage is provided by the bending resistance of the lateral roots on the leeward side; these roots tend to be elliptical or even figure-of-eight-shaped in cross section, ideal at resisting this deformation. However, the vast majority of the anchorage is provided by the sinker roots on the windward side of the trunk; they strongly resist being pulled upwards out of the soil. Sinker roots are so important that when waterlogging stops them from developing, trees can be very unstable.

LEFT When trees like this oak are uprooted they bring up a surprisingly small root-soil plate through which just a few sinker and lateral roots emerge.

BELOW Anchorage failure in a mature tree. As the tree is pushed over it rotates about a point beneath the trunk but on the leeward side. The leeward laterals are bent until they break, while the tap root and windward sinkers are pulled upwards out of the soil.

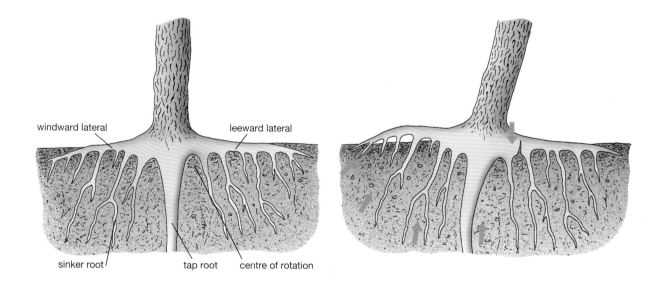

windward lateral   leeward lateral

sinker root   tap root   centre of rotation

Perhaps the most extraordinary anchorage systems are possessed by those tropical rainforest trees that have huge 'buttress roots'. In these trees the lateral roots are particularly shallow to help them exploit the nutrients that are concentrated in just the top few millimetres of soil. Sinker roots are therefore particularly important to anchor these trees; they are widely placed away from the trunk to give them longer lever arms. The buttresses act as angle brackets, transferring forces smoothly down from the trunk to the sinker roots. Without the buttresses the narrow lateral roots would just break.

RIGHT The huge buttress roots of rainforest trees, such as this *Dracontomelon dao* from Java, Indonesia, are needed to strengthen the join between the trunk and the narrow lateral roots.

GROWTH RESPONSES OF TREES

The structure of wood and the architecture of trees are mainly genetically determined. However, trees can fine-tune their mechanical design by detecting their mechanical environment and responding to it with a range of growth responses.

FLAGGING

In areas with extremely strong prevailing winds, such as the tops of mountains or sea cliffs, trees receive forces predominantly from one direction. An involuntary growth response called flagging results. The leaves on the windward side are killed by wind-borne particles and the windward branches are bent gradually leeward by the constant force. The result is that the foliage points mostly downwind of the trunk, which itself leans away from the wind. This makes the tree much more streamlined, reducing the wind forces to which it is subjected. In the most exposed

LEFT At the tree-line in windswept regions, such as here in the Rocky Mountains, trees take up a windswept 'krummholtz' form, clumping together to form streamlined mounds.

BELOW Strong prevailing winds of over 45 m/s (100 mph) have flagged these limber pine trees, *Pinus flexilis*, making them lean and branch downwind.

areas, the wind also tends to kill off the leading shoot at the top of the tree, so that the only living shoots are the ones that point downwind. The tree seems to become bent sideways. Trees exhibiting the prostrate 'krummholtz' form that results are common near the tree line up mountains and in the subarctic.

THIGMOMORPHOGENESIS

Trees also exhibit adaptive growth responses to the wind in areas where there is no strong prevailing wind direction. These responses are called thigmomorphogenesis. The most obvious response is that trees exposed to strong winds grow shorter than those growing in sheltered areas. If you look at the edge of a wood you will see that the outermost trees are shorter than the rest. Tree height increases further in, so many copses seem to have something of a streamlined shape.

RIGHT The trunk of this sweet chestnut, *Castanea sativa*, shows pronounced spiral grain. The wood fibres twist around the trunk, rather than run straight up it, making it more flexible.

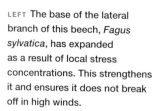

LEFT The base of the lateral branch of this beech, *Fagus sylvatica*, has expanded as a result of local stress concentrations. This strengthens it and ensures it does not break off in high winds.

BELOW LEFT The sides of the wound in this beech, *Fagus sylvatica*, have healed rapidly as a result of local stress, preventing the trunk from being weakened.

Closer examination reveals that the exposed trees also have thicker trunks and thicker structural roots than sheltered ones. The structure of the wood is also altered. Exposed trees have wood in which the cellulose fibres are wound at a larger angle to the axis of the cell. The cells themselves tend to wind around the trunk of the tree rather than running parallel to it, a condition known to foresters as 'spiral grain'. All these changes help make the tree more stable. The reduction in height reduces the drag on the tree, while the thickening of the trunk and roots strengthens them. The changes in the wood, meanwhile, tend to make it more flexible, so the tree can reconfigure more efficiently away from the wind. Trees growing in windy areas even have smaller leaves, and this further reduces drag as well as water loss.

It has been shown that the growth responses of the wood are controlled locally. If a small length of a trunk is bent it will thicken up more than unstressed areas of the trunk, and if it is bent in one plane only it will become elliptical in cross section. In both cases the tree lays down wood where the mechanical stresses are

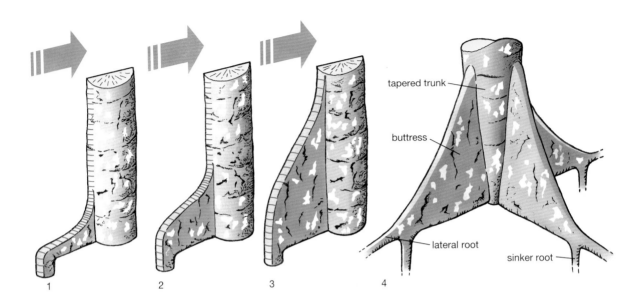

tapered trunk

buttress

lateral root

sinker root

1 2 3 4

ABOVE **Automatic formation of buttress root systems. If a tree is anchored by widely spaced laterals (1), stresses caused by the wind will be concentrated on the top of the root (closely spaced red lines indicate heavily stressed areas). Growth stimulated there results in development of the buttress (2 and 3). 4 shows the mature root system.**

highest. This response is clever, as it ensures that trees only lay down wood where it is actually needed. This facility has been shown to be responsible for many aspects of the shape of trees. It ensures that branches are strongly joined to the trunk by expanding like the bell of a trumpet at their base; stresses are concentrated where the branches join the trunk, and this causes the branch to grow thicker there automatically. It is also the reason why tree wounds heal fastest along their sides – bending stresses along the trunk are diverted around the sides of the wound, and so growth proceeds fastest there. The response also causes lateral roots, which are bent only in the vertical plane, to grow fastest along their tops and bottoms, and so develop into mechanically efficient I-beam shapes. It is even responsible for the growth of the bizarre buttress roots of rainforest trees. When these trees are flexed by the wind, mechanical stresses are concentrated along the tops of the lateral roots; this causes them to grow rapidly upwards, especially at the join with the trunk, and so form buttresses.

The time delay which is inevitable in these growth responses causes problems for us when we grow trees. Cutting a road through a forest or thinning a plantation exposes trees to greater wind forces than they are used to. The result can be catastrophic wind damage before the trees can grow thicker. In urban areas, young trees have traditionally been staked to help support them. Unfortunately, this means that the lower trunk and roots are not mechanically stressed, so they will remain slender and weak. When the stake is eventually removed the trees are therefore extremely vulnerable to damage. Nowadays, arboricultralists advise us to stake trees as near the ground as possible, or bury a wire mesh around the root system to help it to anchor the tree. These precautions minimize the chances of weak areas developing.

REACTION WOOD

Trees react if their trunks are blown over or deflected away from vertical, with growth responses that help them grow vertically again towards the light. The tip of the trunk detects the direction of gravity and automatically bends upwards. The same is also true all the way down the trunk; reaction wood is laid down on one side of the trunk to bend it upwards.

Conifers produce a sort of reaction wood, called compression wood, in which the cellulose microfibrils are orientated at around 45 degrees to the long axis of the cells. This stops the cells from shortening after they are laid down. If a tree is deflected from vertical, conifers produce compression wood on the underside of the trunk, and it tends to push the trunk upwards.

Angiosperm trees produce a very different sort of reaction wood called tension wood, in which the cellulose microfibrils are almost parallel to the long axis of the cell. Cells of this wood tend to shorten even more than normal wood after it is laid down. Angiosperms produce tension wood on the upper side of leaning trunks, and it tends to pull the trunk upwards.

Both compression wood and tension wood are very useful to the trees, but their production has disadvantages for foresters. The two types of wood are both brittle, so planks of wood made from bent trees will not be very strong. The stresses they set up and differences in the shrinkage rates will also tend to warp and split the planks. Hence, misshapen trees have very little commercial value.

BELOW LEFT This cypress, *Cupressus sp.*, has bent a lateral branch upwards to replace its leader.

BELOW RIGHT Different techniques to bend a branch upwards to replace a dead leading shoot. In conifers it is pushed up from below using compression wood. In angiosperms it is pulled up from above using tension wood.

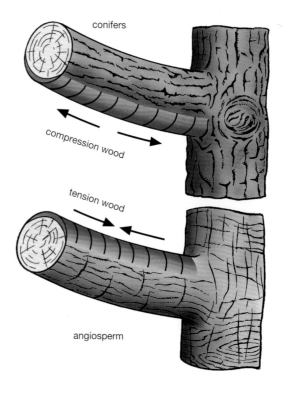

conifers

compression wood

tension wood

angiosperm

CHAPTER 4

Limits to height

LIKE OTHER PLANTS, TREES GROW in a different way from animals, which have a set pattern of development that produces a full-grown adult with a preordained body plan: a set number of legs, for instance. Plants, in contrast, are modular organisms that can produce large numbers of almost autonomous branches, twigs and leaves. They could theoretically keep growing for ever. Yet trees do in one respect show a pattern of growth very similar to animals. They have an S-shaped growth curve, growing slowly when young, much faster at intermediate ages, and slowing down as they mature, until they level off at a maximum height. Why do trees stop getting taller? There must be a genetic component, because some tree species grow taller than others: oaks are bigger than hollies, for example. However, there must also be a connection with the environment, because trees of the same species reach very different heights depending on where they are growing. An oak tree growing in good soil in lowland Britain grows rapidly and can reach up to 30 metres (98 feet) in height; in contrast, one of the gnarled oaks of Wistman's wood on upland Dartmoor, in southwest England, grows slowly and reaches only around 10 metres (33 feet). Similarly, Sitka spruce trees grow up to 60 metres (196 feet) tall in California, but trees of the same species growing in Alaska rarely reach 10 metres (33 feet). Poor conditions reduce both growth rate and maximum height. Several theories have been put forward to explain what limits the height of trees.

RESPIRATION HYPOTHESIS

One theory is that trees grow more slowly as they get taller because they have more wood to maintain. So, as trees grow, more of their photosynthetic production has to be diverted to allow the trunk, branches and structural roots to respire; eventually all of the sugars are needed just to maintain them and the tree stops growing. This theory ties in with the correlation between tree growth rates and maximum heights; fast-growing trees produce more sugars and so can maintain a larger trunk. However, recent research has contradicted this theory; wood is so full of dead material that it requires only around 5–12 per cent of the sugars to maintain it, even in large trees. In any case, the theory cannot explain the fact that trunks of mature trees continue to increase in diameter long after they have reached their maximum height.

OPPOSITE A coast redwood, *Sequoia sempervirens*, growing at Chatsworth Hall, Derbyshire, England. In their native California, redwoods are the tallest trees in the world.

RIGHT This oak, *Quercus robur*, growing in lowland Cheshire, UK has grown close to its potential maximum height of 30 metres (100 feet).

RIGHT Oaks, *Quercus robur*, growing in Wistman's wood, Dartmoor, UK, are stunted by low temperatures and poor soils, so they grow to a height of no more than 10 metres (33 feet).

BRISTLECONE PINES – THE WORLD'S OLDEST TREES

One tree that shows particularly clearly the relationship between the quality of a habitat and the maximum height it can grow is the bristlecone pine, *Pinus longaeva*. Bristlecone pines are extremely hardy trees that grow on semi-arid mountains from the Mexican border northwards to Colorado, where their growth is limited by the extreme cold and drought. The tallest trees, which are generally found higher than 3,000 metres (10,000 feet) above sea level, can grow up to 18 metres (60 feet) tall. Further down the mountain, where drought is more severe, trees cease growing after reaching as little as 4–6 metres (13–20 feet), and develop weird, contorted shapes. The amazing thing about these small trees is their longevity. They can live for over 4,500 years and their heartwood does not rot even when most of the tree's root system has been eroded away. The longevity is probably caused by the same factors that limited the growth. Few disease fungi can survive the drought and sub-zero temperatures, and the dense wood the tree produces is very hard to break down. The longevity of these trees has proved extremely useful to scientists from other disciplines. Using living trees and dead trunks, dendrochronologists have been able to produce continuous growth records of the trees for the last 9,000 years. This has allowed climatologists to reconstruct patterns of climate change. Archaeologists have also used the trunks to calibrate carbon dating results on their finds.

ABOVE **Bristlecone pines *Pinus longaeva* growing at the treeline on Mount Evans, Colorado, USA.**

NUTRIENT LIMITATION HYPOTHESIS

A second theory is that tree height is limited by the availability of nutrients. As trees grow they take up nutrients and sequester them in their leaves and woody tissues; this reduces the availability of the nutrients in the soil. As a consequence, larger trees would have to divert more of their biomass to their roots to maintain the nutrient supply and may not get enough to produce new leaves or branches. There is some evidence for this theory; older forests often do have higher root biomass than young ones, and their soils do tend to have lower nitrogen levels. Furthermore, adding more nitrogen to mature woodland can result in the trees resuming their vertical growth. However, the theory does not explain why young trees can grow perfectly well in old forests if they get enough light. Nor does it explain why parkland trees, which have plentiful nutrients and water, eventually stop growing. The theory cannot therefore be a complete explanation.

MATURATION HYPOTHESIS

A third idea is that trees stop growing because their growing points mature, and their rate of cell division decreases. The slow-down is thus genetically programmed. Trees certainly mature; older shoots are less twiggy than young ones and have lower growth rates. These differences are genetic in origin, as they remain even if

RIGHT Coppiced trees like this hazel (hazelnut), *Corylus avellana*, grow rapidly even when they are hundreds of years old, showing that trees do not stop growing simply because they mature.

a mature shoot is grafted onto a juvenile tree's rootstock. However, the transition to maturity occurs well before height growth slows down. In any case, shoots grow extremely rapidly from the bases of coppiced trees, even when they are several hundred years old. This shows that trees do not age and lose their vigour in the way the hypothesis suggests.

HYDRAULIC LIMITATION HYPOTHESIS

The theory that is best supported by the scientific evidence is that height is limited by the supply of water to the leaves. In taller trees there is higher resistance to water flow because it has further to go up the trunk and because it has to be raised higher against gravity. Because water is dragged up trees under tension, the taller the tree the higher the tension has to be; air bubbles are therefore more likely to form in the xylem vessels, causing loss of their conducting ability. To prevent this, taller trees have to close their stomata earlier in the day or earlier in a drought; this limits photosynthesis, so slowing tree growth.

All aspects of this theory have been verified by experiment. Taller trees do indeed have greater hydraulic resistance and water is under greater tension in their trunks; their stomata do close earlier in the day, and tall forests do have reduced rates of water loss and photosynthesis as a result. What is more, the idea also helps explain three other facts. First, it explains why old trees have flattened tops with gnarled, slow-growing branches and twigs, whereas further down the canopy, growth is more rapid. This occurs because photosynthesis and new growth are most severely reduced at the very topmost twigs, where the hydraulic resistance is greatest. Second, it can also explain why trees growing in nutrient-poor habitats or in cold or dry conditions have a lower maximum height than trees of the same species that are growing in good conditions. These factors will reduce the growth rate of the tree, which will therefore lay down less new wood each year; the result will be a reduction in hydraulic conductivity, particularly since slow-growing trees lay down relatively more 'late wood', which has narrower cells. Future growth is consequently reduced and height is limited.

LEFT The giant redwood, *Sequoiadendron giganteum*, from the western slopes of the Sierra Nevada, California. Growing up to 96 metres (315 feet), it is slightly shorter than the Californian (coast) redwood, *Sequoia sempervirens*, which grows up to 102 metres (335 feet).

Finally, the theory can help us explain why the two tallest species of tree in the world are the Californian (coast) redwood, *Sequoia sempervirens*, and the *Eucalyptus* trees of the Australian temperate rainforest. The climate is ideal for year-round growth in both areas: the eucalypt forest is relatively warm and has plentiful rainfall all through the year; in coastal California there is abundant winter rainfall, while in summer the trees obtain their moisture from the fogs that move in from the sea. Both areas also have fertile soils.

CHAPTER 5

Survival strategies

IN THE PREVIOUS CHAPTERS WE LOOKED at trees as if they were all more or less the same; the main differences we noted were those between conifers and angiosperm trees. However, this is far from being the whole story. The 80,000 or so species of trees vary extensively in almost all their characteristics. They come in many different sizes and shapes; they branch in very different ways; they have very different leaves, flowers and seeds; even their wood varies a lot in colour and density. Botanists have long been fascinated with tree diversity, but it is only recently that we have begun to understand why there are so many species of trees and why they are so different. Partly it is because different tree species are adapted to live in different climates or root into particular types of soil. However, many species of trees can coexist in forests where the physical environment is more or less constant. Different continents also have distinctive tree floras even though they may have similar climates. These facts can only be explained if we consider the biological interactions between trees, and also their evolutionary history.

LEAF ARRANGEMENT AND LIGHT CAPTURE

We have already seen that trees are excellent competitors for light, capable of outgrowing and shading out most other plants. Taking the argument to its logical conclusion, one might expect that in a forest one species of tree would have outcompeted all others for light. The end result of competition would therefore be a single 'supertree'. So why can so many species of tree coexist in a forest? The reason is that different tree species have developed contrasting survival strategies. One crucial difference between them is the way in which they organize their leaves to capture light for photosynthesis.

What is the best leaf arrangement to capture the most light and to photosynthesize fastest? The answer is that it depends on how much light there is. The most obvious way to capture the most light is to produce a single umbrella-like 'monolayer' of leaves. This will intercept all of the available light and cast dense shade. The problem is that the leaves cannot make use of all the light energy

OPPOSITE **In most trees the branches hold up a series of layers of leaves which successively capture the sunlight, as can be clearly seen in this Japanese maple,** *Acer palmatum.*

RIGHT The way in which the net photosynthesis of a leaf varies with the light it receives. Photosynthesis speeds up as the light level rises but levels off above about 20 per cent full sunlight.

BELOW RIGHT The ways in which the net photosynthesis of monolayer and multilayer trees vary with the light they receive. In sunny situations multilayer trees photosynthesize faster, but the reverse is true at low light levels.

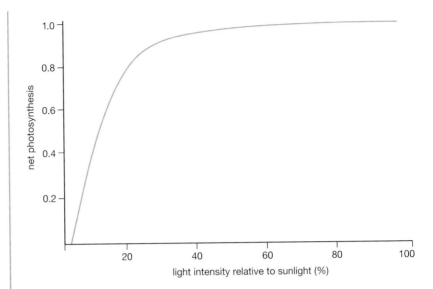

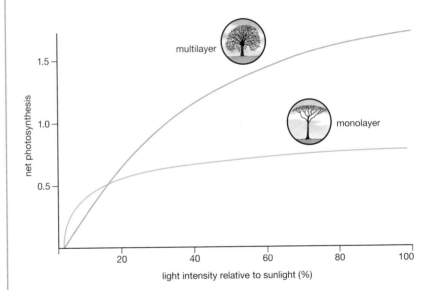

that hits them. The graph at the top of the page shows how the rate of net photosynthesis of a typical leaf depends on the intensity of light that strikes it. Below about two per cent of full sunlight the leaf photosynthesizes so slowly that the energy it produces is outweighed by the energy used to maintain it. As the light levels increase so does net photosynthesis, but only up to about 20 per cent of full sunlight. At this point the leaf's photosynthetic apparatus is saturated, and increasing the light further does not increase the amount of sugars the leaf produces.

In full sunlight it is better for a tree to have a bigger leaf area, but to arrange it into a deep 'multilayer' canopy. Each layer is so diffuse that it lets light down to the layers below. Because leaves photosynthesize at their highest rate even at 20 per

LEFT Beech, *Fagus sylvatica*, a typical monolayer tree, has only a few layers of branches supporting single flat plates of leaves that intercept most of the light. This means the trees cast deep shade, but grow only slowly.

cent full sunlight, the lower, partly shaded, layers can still photosynthesize rapidly; the multilayer canopy, with its higher total leaf area, therefore produces more sugars than the monolayer canopy. Trees with the multilayer leaf arrangement will be able to grow faster in full sunlight. However, there are two disadvantages to the multilayer arrangement. First, if a multilayer tree is put into lower light levels, its lower leaf layers, already shaded as they are by the upper leaves, receive so little light they become a liability to the tree. Therefore, in shade a multilayer

RIGHT The ways in which the light levels beneath the canopies of monolayer and multilayer trees vary with height. Below a monolayer canopy it is much darker.

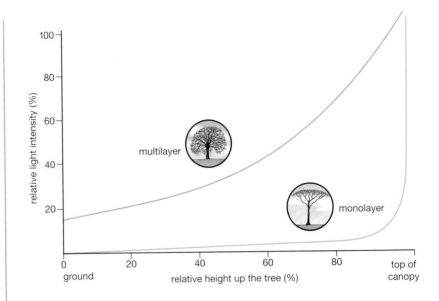

BELOW In birch, *Betula pendula*, a typical multilayer tree, each branch in the deep crown holds up large numbers of randomly orientated leaves, which let a lot of light through.

tree actually grows more slowly than a monolayer tree. Second, in full sunlight, the diffuse leaves of a multilayer tree will let enough light through to allow a monolayer tree to grow underneath. Eventually, the monolayer tree will grow upwards through the multilayer canopy and shade it out.

The important thing to note is that, just as in many other areas of biology, there is no one solution that is best in all situations. Trees that grow well in full sunlight grow poorly in shade, and vice versa. As we shall see, different tree species tend to have leaf arrangements that suit them to the particular ecological niches they take up.

BELOW Autumn in a beech-maple forest in Warren Woods, Michigan, USA. In this late successional wood little light penetrates to the bare forest floor.

ECOLOGICAL NICHES OF TREES

Trees that fill four main ecological niches have been identified: emergent trees, canopy trees, understory trees and pioneer trees. They differ in the arrangement of their leaves, as we might expect, but there are also other considerations to take into account as well as light capture. Consequently, the four types of trees also tend to differ in other characteristics.

EMERGENT TREES

However a tree's leaves are arranged, the most obvious way it can compete for light is to grow taller than other trees. Tropical rainforests contain emergent trees that do just that; seen from above they project up above the main tree canopy layer like hillocks over a plain. One might think that they would be able to shade out the canopy trees that grow below them. However, they simply cannot produce a dense enough canopy of leaves to do this, as several factors limit the area of leaves they can hold up. First, they are so tall that the hydraulic resistance of their trunks limits water transport; consequently they can only supply relatively few leaves. Second, because the crown emerges from the canopy it is more exposed to the wind; therefore these trees have to develop characteristics that increase their stability, and one obvious way is to have fewer, smaller leaves.

Because they are at the limits of tree growth, emergent trees share several characteristics: they have open, sparse crowns; they usually have pinnate or lobed leaves that can roll up into streamlined tubes in the wind and so reduce drag; and, finally, they tend to have flexible, resilient wood in their slender trunks and a strong anchorage system. Examples of emergent trees include the 80-metre (262-foot)-high *Koompassia* trees of Southeast Asia.

CANOPY TREES

The greatest biomass in most forests is held within the canopy trees, whose crowns form a more or less solid layer of leaves when seen from above. They have fewer problems than the taller emergent trees in supplying their leaves with water and standing upright, so they can hold up a much denser crown. Their leaves are usually held in something close to the monolayer arrangement with just a few dense layers of leaves, so although the trees are relatively slow-growing, they cast relatively deep shade. This stops any but the best shade-adapted trees from growing beneath them.

Canopy trees share several other characteristics that help them survive and hold their positions for hundreds of years. They have relatively dense wood that is resistant to the

BELOW Emergent *Koompassia excelsa* trees tower over the canopy of a tropical rainforest in Danum Valley, Borneo.

mechanical forces they have to withstand. The wood is also heavily impregnated with chemicals such as terpenes and phenolics, which often give it a dark colour. These chemicals repel boring insects and fungi that would otherwise destroy the wood. Typical canopy trees include the oaks and beeches of Europe and North America, the spruces and pines of the northern temperate areas and subarctic forests, and the giant trees of tropical rainforests, which yield timbers such as mahogany, teak and meranti.

UNDERSTORY TREES

Canopy trees intercept the great majority of light, but enough gets through to support a third layer of smaller understory trees. Understory trees are extremely well adapted to survive and reproduce in deep shade; they produce a very shallow monolayer canopy, consisting of just a single umbrella of leaves, which captures

LEFT An understory of shade-tolerant Pacific dogwood trees, *Cornus nuttallii*, growing beneath a canopy of giant sequoia, *Sequoiadendron giganteum*, King's Canyon National park, California, USA.

almost all the remaining light. This allows them to survive, but they grow so slowly and hence have to live so long, that their wood needs extreme adaptations to defend it from disease; it tends to be even denser and even darker than the wood of canopy trees. Typical understory trees include the hazels (hazelnuts) and dogwoods of Europe and North America. The ebony trees of tropical rainforests have heartwood which is so dense that it sinks in water and so full of defence chemicals that it is black.

THE CLIMAX FOREST

Many old forests contain all three of these types of trees: the emergent, canopy and understory trees. In this 'climax' community, they make up a canopy with three layers that efficiently harvests the light. Together they reduce light levels on the forest floor to around only two per cent of full sunlight, and it is so dark that few tree seedlings can grow. Old forests therefore seem to be extremely stable environments; the existing trees keep all other trees from invading. However, no tree can live for ever, and sooner or later the forest canopy will be broken; a single tree may die or a larger gap may be produced when a hurricane or storm blows over or damages many trees. The spaces that are opened up are colonized, at least for a short time, by a fourth type of tree, the pioneer tree.

PIONEER TREES

Pioneer trees are adapted to invade, grow rapidly and reproduce in forest gaps, and they tend to share several characteristics. Many have a deep 'multilayer' canopy with large numbers of randomly placed leaves that allow rapid photosynthesis. Many also reduce their support costs by producing light wood, or a hollow trunk. They economize on expensive defence chemicals, and so they have light-coloured wood that is extremely vulnerable to fungal attack. This does not matter, because they mature so fast

BELOW The shiny, peeling bark of birches is ideally adapted to protect the tissue beneath from temperature changes. It reflects most of the sunlight to prevent the outside heating up and is a good insulator of heat.

BELOW RIGHT A mid-successional forest, such as this forest at Finger Lakes, Ithaca, USA, is typically a mix of intermediate species, such as maples, as well as ageing pioneer and young climax trees.

that they do not have to live very long. Typical pioneer trees include the birches and poplars of northern Europe and North America, many pine trees, and the balsa trees of the Amazonian rainforest. Several other species such as sycamores, maples and hickories, seem to be intermediate in ecology and form between pioneer and canopy trees. They are slower growing than pioneer trees but cast somewhat denser shade because they have canopies that are intermediate between the multilayer and monolayer design.

FOREST ECOLOGY AND SECONDARY SUCCESSION

A typical area of forest is in fact a bit of a patchwork. It consists of areas of climax forest with gaps of various ages filled with pioneer trees. Beneath these pioneers, young emergent, canopy and understory trees are growing and gradually taking over again. If a fire destroys an area of forest or if it is cut down, it will gradually regrow in a predictable pattern that is called secondary succession. The first trees to reinvade and to grow upwards will be the pioneer trees. The intermediate species and the climax trees will take longer to recolonize and will grow more slowly under the pioneer trees; eventually, however, they will grow through the pioneer trees, shade them out and take over. After a matter of 100 years the forest will be dominated by intermediate species, and after 200 years or so the climax forest will have been re-established and will be dominated by canopy and understory trees.

CHAPTER 6

Trees in different climates

FORESTS IN DIFFERENT PARTS OF THE WORLD contain trees that fill much the same niches. However, the actual species of trees growing in the tropics and temperate areas look very different. The trees even come from quite different plant groups; angiosperm trees dominate the tropics and wet temperate areas, whereas conifers dominate the subarctic and drier temperate areas. This chapter examines how climate affects the distribution of trees, and how it affects the characteristics of the trees and forests that are found in different parts of the world.

CLIMATE AND PLANT GROWTH

The Earth's climate is driven by the warming action of the Sun's rays. Although climate is also affected by sea currents, and the position of the continents and mountain ranges, the most important single factor that controls it is latitude. The strong rays of the Sun at the equator set up three major convection cells. Seasonality is caused because the convection cells move northwards and southwards over the course of a year as the tilt of the Earth relative to the Sun changes.

The climate is perfect for plant growth all year round only in a slim band within about five degrees of the equator called the intertropical convergence zone. Here the bright morning sunshine heats up the vegetation and evaporates water from it, causing the upward convection of wet air. As it rises, this air forms clouds and produces rain in the afternoon and evening. Therefore the climate is warm, wet and windless throughout the year.

Away from the equator, at latitudes of 5–25 degrees, plant growth is greatly restricted by a dry season that occurs when the Earth is tilted away from the Sun. At this time the region is directly below the area where cool air falls to earth. The air heats up as it falls and any clouds evaporate. Plant growth is therefore only possible in these regions during the summer monsoon, when the region is tilted towards the Sun. At latitudes of 25–35 degrees the land is permanently in the region where cool air is falling to earth; consequently, the climate is dry and unsuitable for plant growth all year round, and deserts form.

OPPOSITE Canary Island pine, *Pinus canariensis*, growing on volcanic rock in La Palma, Canary Islands. The trees use their long needles to strip water droplets from the clouds that are brought to the subtropical islands by the northeast trade winds.

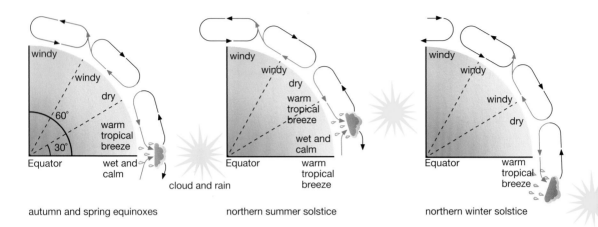

windy

windy

dry

60°

warm
tropical
breeze

30°

Equator   wet and
                 calm

cloud and rain

autumn and spring equinoxes

windy

windy

dry

warm
tropical
breeze

wet and
calm

Equator   warm
                 tropical
                 breeze

northern summer solstice

windy

windy

windy

dry

Equator   warm
                 tropical
                 breeze

northern winter solstice

ABOVE Air currents around the northern hemisphere driven by the warming action of the Sun. During equinoxes these produce wet and calm conditions around the equator, dry conditions at 30 degrees latitude and increasingly windy conditions further north. The three convection cells are displaced northwards in summer and southwards in winter, resulting in seasonality.

Above latitudes of 35 degrees the growing conditions improve, particularly on west-facing coasts, where in winter the land is watered by poleward-bound winds that have picked up more water from the sea. At latitudes of around 40 degrees the west-facing coasts of the Mediterranean, California, Chile and Australia pick up enough winter rainfall to allow reasonable plant growth, even though they are still very dry in summer. Their climates are known as Mediterranean climates.

At latitudes above around 45 degrees, the climate becomes damp and windy all through the year, but cold restricts plant growth during the winter. The length of the growing season falls at higher and higher latitudes as winter lengthens; above 70 degrees winter lasts all year round and no plants can grow.

LIMITS TO TREE DISTRIBUTION

Because trees are taller than other plants and have a shoot system that is permanently exposed above ground, they are unable to survive in the extreme climates of some other plants. Naturally, forests would be restricted to only 40 per cent of the dry land, and to several distinct areas. The central tropics are covered by tropical rainforests, while outer tropical areas have seasonal monsoon forests. The wetter temperate regions are home to three very different sorts of forest: coastal areas around 40 degrees latitude are covered by Mediterranean forests, while other temperate and subpolar areas are covered by temperate deciduous woodland and boreal forest.

Areas that can support other plants but not a complete cover of forest include the heavily seasonal savannahs of the outer tropics and prairies or steppes of the drier temperate regions. Both areas are dominated by grasses, as tree growth is restricted by fire, grazing by herbivores and lack of water. In the subarctic and

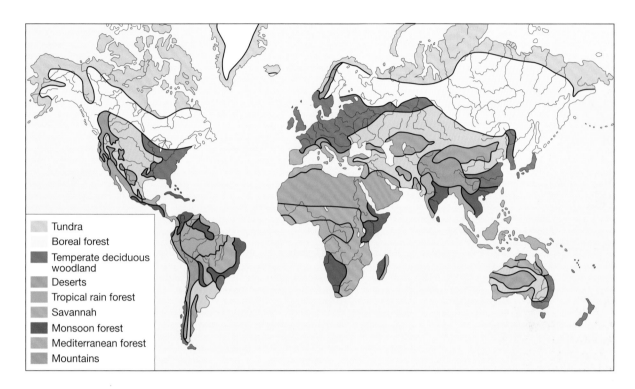

Tundra
Boreal forest
Temperate deciduous woodland
Deserts
Tropical rain forest
Savannah
Monsoon forest
Mediterranean forest
Mountains

arctic tundra the summer is too short for trees to reach a reasonable size and the winter too cold for their foliage to survive. Here the vegetation is mainly made up of shrubs and perennial herbs. In deserts it is so dry all year round that only a few scattered plants can survive.

ABOVE The distribution of different vegetation types is controlled by climate and consequently linked to latitude, with rainforests and monsoon forests near the equator, Mediterranean and deciduous woodland in mid latitudes, and boreal forest in the subarctic.

TROPICAL RAINFORESTS

Tropical rainforests, which have fascinated biologists for hundreds of years, are found in three main areas of the central tropics: Amazonia, central Africa and Southeast Asia. They are the forests with the fastest-growing trees and contain a bewildering diversity of organisms. The number of tree species alone is staggering; one survey of the Amazonian rainforests found well over 500 species of large trees within a single hectare (2.5 acres). The giant trees are also the backbone of an ecosystem that supports a wide range of other plants as well as huge numbers of insects and other animals. As we shall see, the peculiarities and sheer bounty of tropical rainforests are due to their warm, wet and windless climate.

ADVANTAGES OF THE ANGIOSPERMS

Tropical rainforests are dominated by angiosperm trees, because these have a competitive advantage over conifers in the equable climate. In the tropics there is so little wind that the diameter of a tree's trunk is restricted mainly by the need to supply the leaves with water. Angiosperms do not need such thick trunks as conifers because they are plumbed with more efficient xylem vessels; they can therefore grow taller more quickly. The angiosperms' animal-pollinated flowers give them an added advantage because in calm conditions animals are far better than wind at transferring pollen. It is not surprising, therefore, that there are only a very few conifers in rainforests; the few conifers that do exist, such as the broadleaved conifer *Agatbis*, are restricted to windier upland areas.

FORM OF RAINFOREST TREES

The climate is so good for tree growth that primary rainforests have climax trees that fill all three of the major ecological niches: emergents, canopy trees and understory trees. Despite their great diversity, though, most rainforest trees actually look very similar to each other. They have tall, slender trunks supported with root buttresses, and the trunks support crowns that are small and bear thick, evergreen leaves. The leaves themselves are also very uniform, being large and oval-shaped, like laurels, but with an extended 'drip tip'; the tip helps the leaf to shed rain water as a stream of small droplets, which, recent studies have shown, protects the soil below from excessive erosion. Most of the trees produce flowers and fruit in their canopy; the flowers mostly attract insects for pollination, while the fruit is eaten by birds and monkeys, which inadvertently disperse the seeds. Some trees, though, display 'cauliflory'; they have flower stalks that grow directly out of the trunk. The trunk can support large bat-pollinated flowers and huge fruits that attract ground-dwelling animals.

BELOW The leaf of a tropical rainforest tree, showing the extended 'drip tip', which helps shed tiny water droplets.

BELOW RIGHT Cauliflory in a tropical rainforest fig tree. The flowers and fruit grow directly out of the branches and trunk rather than at the tips of shoots.

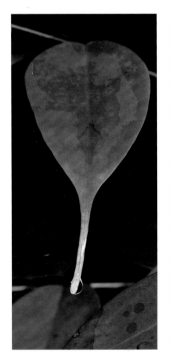

The climate also influences two other less immediately visible attributes of rainforest trees. They only have thin bark because there is no risk of frost. The equable climate is ideal for fungal growth, however, so rainforest trees have to defend their wood against rotting by incorporating defence chemicals, which tend to stain the wood dark.

LIANAS AND EPIPHYTES

The complex rainforest canopy is so efficient at harvesting light that very few plants can survive on the forest floor. There are only a few tree seedlings marking time until a tree falls over and a light gap opens up. However, the warm moist conditions allow two other sorts of plants to take advantage of the mature trees. Epiphytic plants such as orchids, bromeliads and ferns perch high up on their branches, while woody lianas climb on or around their trunks to reach the light. Most of these plants are not too harmful to the trees, but strangler figs are adapted to kill their host and take over its position. They start life as epiphytes, before developing a mass of aerial roots that grow down the trunk to the forest floor. The roots gradually grow thicker and join up, eventually strangling the host tree. As the host dies and its trunk rots away, the aerial roots carry on growing, producing a kind of lattice-work trunk that supports the crown of the usurper.

ABOVE Rainforest lianas, such as this in Sabah, Malaysia, have such wide vessels that they can transport water to their leaves even up their very narrow stems.

LEFT Epiphytic bromeliad, such as this from the cloudforests of Chiapas, Mexico store rainwater for photosynthesis in their 'tank' of overlapping leaves.

RAINFOREST PIONEER TREES AND SUCCESSION

When a gap is opened up in the rainforest it is rapidly filled up by a tangle of herbaceous plants and climbers. At first glance it would seem impossible for young trees to break through this tangle and re-form the forest canopy; the climbers would simply smother them. Fortunately for the rainforest, their pioneer trees have evolved several ingenious methods of preventing this from happening. Most have bizarre umbrella-like shoot systems; they have a single thick trunk, from the top of which a few radiating branches hold up a canopy of huge leaves. The thick slippery trunks provide no grip for twining plants, while the branches are held up well out of reach of grasping tendrils. Many pioneers also enlist the help of ants to remove the surrounding vegetation. The trees shelter the ants within their hollow twigs and provide them with nectar; in return the ants bite through the growing tips of any climber that touches the tree. They also kill and eat small herbivorous insects which might otherwise attack the pioneer tree.

The downside of the shallow canopies of tropical pioneers is that their photosynthesis is limited. However, the trees make the most of what sugars they do produce by having very lightweight trunks. Balsa trees, for instance, have wood with a density of only around 0.15 grams per cubic centimetre (9 pounds per cubic foot), one-quarter that of most trees. The trunks of *Cecropia* trees are even lighter, as they are hollow with thin walls and strengthening bulkheads. Rainforest

BELOW *Cecropia*, a rainforest pioneer tree from the New World tropics, with a slender trunk holding up a typically simple crown of huge leaves.

pioneers grow very rapidly and flower after just 10–20 years, but they provide perfect sheltered conditions for climax trees to grow below them. They therefore quickly lose their place in the canopy. This makes efficient reproduction essential. Many pioneer trees have large, bat-pollinated flowers, and they produce huge seeds that can survive for years in the soil. They remain dormant until another gap opens up in the canopy and the light stimulates them to germinate.

CAUSES OF RAINFOREST DIVERSITY

No one is really sure why rainforests have such huge biodiversity. The fact that conditions are ideal for plant growth obviously helps, but this does not explain why there are hundreds of species of trees; one might expect that competition between trees would produce just a few winners. Several other factors must also be responsible. First, rainforests are very old, having existed for tens of millions of years, so there has been a long time for new species to evolve. Second, new species might evolve more rapidly because the trees are animal-pollinated; separated populations of trees could be visited by different species of insect, and this might lead to evolution of the form of their flowers and cause new species to arise. Third, a single species would be prevented from dominating because the climate is also ideal for insect herbivores, fungi and viruses, which would evolve rapidly and target the dominant tree. There would therefore be strong selection pressure for new species of trees to evolve and develop novel defensive chemicals to repel their attack. It is no surprise that rainforest plants are a potentially limitless source of drugs; many plants are already used medicinally by the indigenous peoples, and drug companies are setting up large research programmes to look for more.

ABOVE Close-up of the trunk and leaf stalks of *Cecropia*, and the ants that protect it from herbivores and climbers. The tree shelters the ants within its hollow twigs. The ants feed on the secretions that exude from the pads at the bases of the leaf stalks.

THE PRESENT AND FUTURE OF RAINFORESTS

Huge areas of rainforests have been damaged or destroyed in the past century by forest clearance for agriculture or by logging. Obviously, clearing the land totally destroys the rainforest ecosystem. Logging, however, has a rather more complex effect depending on how the operation is carried out. If the forest is not burned, the vegetation can recover, just as when a gap occurs in the canopy. Climbing herbs will grow first, before being replaced by pioneer trees, and then eventually by climax trees, which form a secondary forest. The long-term effect might not therefore be that bad. The problem is that the 'gaps' created by humans are unnaturally large, logging is often repeated far too often, and soil can be badly damaged when the trees are extracted. All this means that the secondary forests are not nearly so diverse as the original primary ones. Our hope for the future of rainforests is that some areas of primary forest might be left intact, while methods of more sustainably harvesting the secondary forests can be developed.

MONSOON FORESTS

Monsoon forests are for the most part very similar to tropical rainforests. The trees are usually tall, most of them are angiosperms, and they have thick, laurel-like leaves. There are differences, however, which are related to the selection pressures imposed by the dry season.

THE TREES

The biggest difference is that the trees are deciduous; they shed foliage at the end of the wet season along a specially weakened layer of cells called the abscission layer. This prevents undue water loss during the dry season, but it means that the trees lose a lot of material each year and must produce a new canopy of leaves at the start of the wet season before they can start to grow again. The trees reduce losses of nutrients by transporting organic compounds and chlorophyll from the leaves back into the woody tissue. They speed up the development of the new canopy by forming drought-resistant buds at the end of the wet season. These contain pre-formed shoots that are ready to expand to form a new canopy as soon as the rains return. However, during the dry season they are protected by being encased in drought-resistant scales. The trees' trunks are also protected from extremes of temperature in the dry season by being encased in thicker bark.

BELOW Torrential monsoon rain drives rapid tree growth in the tropics, but only occurs for part of the year in monsoon forests.

OTHER PLANTS

Because the trees are deciduous, a lot of light reaches the floor of monsoon forests during the dry season. This allows a dense undergrowth of drought-resistant herbs and shrubs, especially grasses, to cover the ground. On the other hand, the dry season effectively excludes two groups of plants that are very successful in rainforests: lianas and epiphytes. Lianas are too prone to embolisms of their wide xylem vessels during drought, while the only epiphytes that can survive the prolonged desiccation are lichens and a few ferns.

TREES IN TEMPERATE AREAS

In the colder, windier temperate regions, angiosperm trees do not necessarily have a competitive advantage over conifers. Their wide xylem vessels are more likely to be blocked by embolisms, especially at higher latitudes, and this reduces their water-conducting advantage. Furthermore, the efficiency of insect pollination falls at higher latitudes because there are fewer flying insects, whereas the efficiency of wind pollination increases. Therefore, both conifers and angiosperms thrive in temperate areas; the precise local conditions determine which group dominates.

EVERGREEN OAKS AND DECIDUOUS CONIFERS

In temperate areas most conifers are evergreen, whereas most angiosperm trees are deciduous. However, this is not a hard-and-fast rule, especially in other parts of the world; both sorts of trees adapt their pattern of leaf loss and replacement to their local environment. In the tropics most angiosperm trees are evergreen because of the lack of seasonality. Similarly, many angiosperms from Mediterranean regions, such as olives or evergreen oaks, retain their leaves throughout the year because it would be unprofitable to grow new leaves for each short winter growing season. Even in temperate areas some angiosperms are evergreen, notably understory trees such as hollies and boxes (boxwoods); they get too little light to replace their leaves every year. Conversely, conifers such as swamp cypresses (bald cypresses) are deciduous because they live in areas with a long growing season, where short-lived rapidly photosynthesizing leaves are more profitable. So, too, is the more familiar larch; this tree is probably deciduous because in its native Siberia the winter is just too harsh for evergreen leaves to survive.

DECIDUOUS OR EVERGREEN?

Trees can use one of two strategies to survive the seasonal cold and drought of temperate areas. The technique most often used by angiosperms is to be deciduous; they shed their leaves at the end of the growing season, and so limit water loss. The technique used by most conifers is to be evergreen; they develop more drought-tolerant evergreen foliage, which has fewer stomata and a thicker cuticle. Both strategies have advantages and disadvantages. On the one hand, evergreen trees do not lose anywhere near as many leaves each year as deciduous ones, so they do not need to divert so much energy into producing new foliage. On the other hand, their leaves photosynthesize more slowly than deciduous leaves because the adaptations they make to reduce water loss limit their uptake of carbon dioxide. For this reason, the best strategy to use depends on the precise growing conditions. If there is a long, productive growing season, deciduous leaves will repay their greater investment. In contrast, if the growing season is short it may be better to have evergreen leaves.

MEDITERRANEAN FORESTS

In the coastal areas of the Mediterranean and California in the northern hemisphere and Chile and Western Australia in the south, tree growth is limited by summer drought to the short, damp winters. The brevity of the growing season has two effects on the Mediterranean forests that grow there. First, almost all the trees are evergreens, although by no means all are conifers. In the Mediterranean itself, for instance, there are conifer forests of Aleppo and umbrella pines, cypresses and cedars, but angiosperms such as evergreen oaks, carob trees and olives also thrive.

BELOW The Calabrian pine, *Pinus brutia*, seen here in Cyprus, is a major forest tree in the eastern Mediterranean.

Second, the trees tend to grow slowly and are often much shorter than the trees of tropical forests. The trees have even thicker bark than the trees of monsoon forests, which protects the phloem from extremes of heat and cold and from forest fires; it is no accident that cork oaks come from the Iberian Peninsula. Because the trees are so short, there is just a single canopy layer, below which there is a ground layer of shrubs such as heathers and rock roses.

Much of the vegetation of California resembles that of the Mediterranean; there are extensive forests of relatively short evergreen oaks, pines and cypresses with an understory of evergreen shrubs. However, California is also home to the tallest tree in the world, the Californian (coast) redwood, *Sequoia sempervirens*. These trees grow to over 102 metres (335 feet) and can live for 2,000 years. The existence of such tall trees in such a drought-stricken area seems puzzling. However, Californian (coast) redwoods are in fact confined to a narrow coastal belt where there is a high winter rainfall of 100–200 centimetres (40–80 inches). This is augmented in summer by coastal fogs that condense on the foliage and drip to the floor, to add a further 25 centimetres (10 inches) of precipitation. The trees' roots are well placed to exploit this extra source of water; they grow close to the surface and spread widely. With such a good year-round water supply, redwoods are among the fastest-growing of trees and can reach

their huge size within a century. Another reason for their rapid growth is that, unusually for long-lived trees, redwoods have a very deep multilayer canopy, which drives their growth to record heights. They eventually become so tall that they cannot supply enough water to their tops of their crowns to expand their needles, which end up as short stubby scales. The result is that the trees are not only fast-growing but produce very dense shade. This is why redwood forests have so little undergrowth beneath them. There is only one disadvantage; redwood seedlings find it hard to grow under their own canopy, and the cones have to wait until the forest is destroyed by fire before they open and the seedlings sprout upwards. Fires of sufficient intensity occur only very rarely, because redwoods are protected by extremely thick bark and contain very little resin. This is why the trees in so many redwood forests are all very old, and the same age.

Mediterranean forests are also found in Chile and Australia, but for reasons we will examine later in the chapter about southern hemisphere trees, they contain quite different tree species.

ABOVE The largest living tree in the world, 'General Sherman', a giant redwood of the species *Sequoiadendron giganteum* from California's Sierra Nevada, USA is estimated to be over 80 metres (265 feet) tall and to weigh at least 2,500 tonnes (2,550 tons).

TEMPERATE BROADLEAVED ANGIOSPERM FORESTS

In central temperate areas trees can grow for around 6 months of the year, and the long wet summers have mean temperatures of over 10°C (50°F). With such a long and productive growing season, deciduous trees outcompete evergreens, and with the mild winters angiosperms outcompete conifers. It is no surprise, therefore, that western Europe, the northeastern USA and southern Japan were in prehistoric times covered in deciduous woodland that was composed mostly of angiosperm trees. Unfortunately, only a remnant of this forest, usually heavily managed, remains because the rest has been cleared for agriculture. In Europe the best place to see apparently untouched broadleaved angiosperm forest is in the Bialowieza forest of Poland and Belarus (Byelorussia).

The long summer growing season enables the trees to grow rapidly and quite tall. The forest ecosystem is also quite complex, supporting canopy trees such as beeches, oaks and limes (basswoods), and understory trees such as hornbeams, hazels (hazelnuts) and dogwoods. Many of these trees have adapted to the windy conditions by reverting to wind pollination. They flower in early spring before the leaves are produced; the male flowers release their pollen from long dangling catkins and the female flowers catch it on feathery stigmas. Many temperate trees also have lobed or pinnate leaves that can roll up in the wind to reduce drag.

Temperate broadleaved angiosperm forests harbour a distinctive flora of other plant species. Just as in monsoon forests, there are very few lianas, one of the few successful species being the root climber ivy, *Hedera helix*. The only epiphytes are ferns, mosses and lichens, which can cope with dry periods by

RIGHT The large winter buds of broadleaved trees such as this horse chestnut, *Aesculus hippocastanum*, contain pre-formed leaves. This allows the canopy to open and start photosynthesis early in spring.

FAR RIGHT A catkin, or male flower, of a hazel (hazelnut), *Corylus avellana*. It releases its pollen as it dangles in the wind.

TOP LEFT Woodland bulbs, such as these bluebells, *Hyacinthoides non-scripta*, shoot up and perform their photosynthesis before the canopy above them opens. This allows them to carpet the forest floor with spring flowers.

ABOVE Mistletoe, *Viscum album*, a hemiparasite that uses its roots to obtain water and nutrients from the xylem sap of its host tree.

LEFT The roots of the monsoon rainforest kapong tree, *Tetrameles nudiflora*, encroaching over the Angkor temples in Cambodia. The tree is itself being encroached by the narrower roots of a strangling fig, which may eventually kill it and take over its position in the forest.

going into suspended animation. One angiosperm that looks like an epiphyte – mistletoe – is actually a hemiparasite, a plant that produces its own sugars by photosynthesis, but that plugs its roots into the wood of the tree in order to obtain water. There are also a few understory shrubs and ferns. The most successful and characteristic woodland plants though, are perennial herbs, especially monocotolydons. The trilliums of North America and the bluebells and wild garlic of northern Europe emerge and flower in early spring, growing and flowering before the trees above have opened their leaves. These plants are able to grow so quickly in spring because they carry out all the cell divisions they need to form their new shoot systems during the previous autumn. They store the pre-formed plant in their bulbs and merely expand them in spring, carpeting the forest floor with colour.

The pioneer trees of deciduous forests, such as birches and poplars, do not have as much difficulty outcompeting herbs as the tropical pioneers, which have to contend with lianas and other climbers. They are therefore quite unlike tropical pioneer trees. They have deep canopies and hold up large numbers of small leaves that are free to wave about in the wind. Their wood is light in colour, as it contains few defence chemicals, and their trunks are protected from being overheated by direct sunlight by being coated in pale, shiny bark.

Despite their high productivity, deciduous forests, especially those of western Europe, are far less diverse than tropical rainforests. Partly this is because the climate is poorer. However, much of the difference is probably due to historical events. Deciduous forests have been forced to migrate towards the equator and back again several times in the last two million years because the Earth has been gripped by successive ice ages. Of course, the trees do not move themselves! Instead, a forest moves as the seeds disperse and germinate at the extremes of the forest's range. In North America the trees were free to move south because the main mountain chains have a north–south orientation. In Europe, however, southward movement of deciduous forest was blocked by the Pyrenees and Alps, which have an east–west orientation. Far fewer tree species therefore survived in Europe. Even fewer got back to Britain before the land bridge was broken after the end of the last Ice Age. Several species that could survive in Britain were therefore absent before they were introduced by humans: good examples are sycamores and sweet chestnuts, and it has even been suggested that beech is an introduced species.

BOREAL FORESTS

The northern temperate and subarctic regions of Canada, northern Europe and northern Asia, have summers that last for less than six months and have a mean temperature below 10°C (50°F). Here evergreen trees have a competitive advantage, and conifers, with their greater ability to withstand embolisms, are better at surviving than angiosperms. Therefore, the region is covered in coniferous boreal forest or taiga.

Most of the conifers of boreal forests are members of the family Pinaceae – pines, spruces, hemlocks and firs – and possess the characteristic frost-tolerant needles. In the warmer, southernmost forests they can grow into huge trees, below which there is a subcanopy of smaller angiosperm trees such as maples, and a ground layer of shade-tolerant ferns. Apart from these few plants, however, little else grows beneath the trees, partly because their evergreen foliage shades the ground all year round. Also, the thick needles that fall to the floor take so long to break down that they form an acidic layer that inhibits plant germination. An

LEFT A mixed broadleaved/ conifer woodland in Pennsylvania, USA, showing the understorey of dogwoods and the forest floor covered by ferns that survive on the little light that penetrates the canopy.

LEFT In boreal forests, tree seedlings often grow fastest on fallen tree trunks, known as 'nurse trees', because they contain more nutrients than the poor soils.

exception is the Caledonian forest of Scotland. The soil here has very low fertility and can only support a broken canopy of Scots (Scotch) pines and a few smaller birches; the ground below them is covered with heather, bilberries and mosses. Another region with open forest is Siberia, where the particularly harsh winters prevent even the toughest foliage surviving. Here there are forests of deciduous larches and birches.

As you move further northwards and the climate gets colder, the trees grow more and more slowly and become shorter and shorter. Many display special adaptations to prevent themselves being crushed by snow. Instead of having a bushy top, they are conical in shape, with branches and twigs that point downwards; both of these adaptations allow snow to slide off the tree to the floor. Further north still, the trees grow too far apart to produce a complete canopy and the ground between them becomes covered by mosses and lichens. Eventually it becomes too cold for trees of any size to grow, and the ground becomes covered in tundra. Even here, though, willows and birches do grow. They are just too difficult to see because they are often no more than a couple of centimetres (an inch or so) high!

NORTHERN MONTANE FORESTS

In many respects, mountains have a climate that is just like that of the subarctic. They have short summers and long, cold winters. For this reason, mountains tend to be covered in conifer forests that are very similar to the boreal forests of the subarctic. There are pines, spruces and firs, and the higher up the mountain they grow the shorter and more conical they become until they die out at the tree line. The main difference between montane and lowland trees is

RIGHT Trees growing near the tree line like these in the Rocky Mountains, Canada, often have narrow, conical canopies that help them to shed winter snow.

LEFT Scots (Scotch) pine forest at Rothiemurchus, Cairngorms, Scotland, UK, showing the open canopy typical of such forests with its understory of heather.

BELOW Trees get progressively shorter and more wind-sculpted the higher up a mountain they grow, before disappearing altogether at the tree-line, as here in the Rocky Mountains, Colorado, USA.

that they are adapted to very different light regimes and patterns of freezing and thawing. Montane trees tend to use the temperature rise as the snow melts in spring as a reliable indicator of when they can start to grow; they can do this because once the deep winter snow has melted it is unlikely to snow again. In the lowlands, thawing is more likely to occur from time to time, even in winter. Consequently, the time of year is a more reliable indicator of when it is safe to start growing again without suffering frost damage. Lowland trees therefore use the increasing daylength as a cue of when to start growing. The problem for foresters is that trees from different latitudes as well as from different altitudes are adapted to their own local climate. So although the range of Sitka spruce for example extends from California in the south to Alaska in the north, it is only safe to plant trees with local provenance.

CHAPTER 7

Specialist trees

MOST TREE SPECIES ARE ADAPTED to live together in forests. Some, however, are specialists that live in particular habitats: along rivers, in swamps, around the seashore, or even in dry deserts far from other trees. As we shall see, many of these trees exhibit unusual adaptations that give them an advantage in these situations.

OPPOSITE **Grandidier baobab,** *Adansonia grandidieri*, **trees from Western Madagascar. Baobabs live in seasonal forests in Madagascar and South Africa, surviving prolonged arid periods by losing their leaves and storing water in their swollen trunks.**

BELOW **Many riverside trees like this willow,** *Salix alba*, **have leaning trunks that enable them to maximize their light capture.**

RIVERSIDE TREES

The banks of rivers and streams are in many ways ideal for tree growth. The river produces a linear gap in the forest canopy; consequently, riverside trees are well supplied with light, at least on the side facing away from the bank. The river also brings in nutrients and keeps the soil moist. The river does bring problems as well, however. In times of flood the soil will become waterlogged, preventing oxygen getting to the roots. In particularly bad floods, the bank and the tree may even get washed away, or parts of the tree's shoot system may be broken off by the force of the water.

Riverside trees survive flooding by piping air down to the fine roots. They develop large air spaces called aerenchyma within their roots, and these air passages are connected to pores in the trunk called lenticels, to form a sort of air supply line. This technique is very successful, and the floodplain trees of the Amazonian rainforest can survive being submerged to depths of 10 metres (33 feet) for over six months a year. The root systems of riverside trees are massive, and help to stabilize the bank. The root systems need this size and strength, because many species have adaptations of the shoot system that put the roots under greater strain. Often, the trunk is not held vertically, but leans in towards the river, exploiting

the extra light from that direction. It can be seen that this pattern of growth is genetically determined in willows and black poplars, because they lean even if they are grown in open parkland where they are not shaded at all!

Some trees even take advantage of flooding to help them disperse. Willows, such as the European crack willow, *Salix fragilis*, have twigs that readily break off the tree at the specially weakened base; the twigs float downstream and get washed up on a new bank, where they root and produce a new tree.

SWAMP TREES

Trees that live in waterlogged swamps have much the same problems with waterlogging as those that live by rivers, and they solve them in much the same way; most of them have roots with aerenchyma plumbing. However, trees such as the swamp cypress (bald cypress) also use another technique that supplies air nearer to where it is needed. The lateral roots develop special hollow breathing organs called knee roots, or pneumatophores; they grow upwards out of the

BELOW The pneumatophores, or 'knee roots', of swamp cypresses (bald cypresses), *Taxodium distichum*, emerge from the water, helping pipe air to the roots.

water, allowing air in through their lenticels, just like so many mini trunks. This technique allows swamp cypresses to form extensive forests in the flooded plains of the southwest USA, such as the famous Florida everglades.

Many bogs have another feature that inhibits tree growth; they have very low levels of nutrients available, especially nitrogen. Alders improve their ability to colonize bogs and recently melted glacial soils by having a mutualistic relationship with a bacterium called *Frankia*. The bacterium fixes nitrogen using energy obtained from the roots, and the roots in return receive some of the fixed nitrogen. As alders grow they therefore not only start to dry out the soil but also improve its fertility. Alder woodland was an important stage in the colonization of northern Europe and North America at the end of the last ice age, as it improved the soil enough to allow spruce and other conifers to thrive. Nowadays alders are commonly planted around the outskirts of towns, where they help improve the soil of former landfill sites and other restored industrial areas.

ABOVE Alders, *Alnus glutinosa*, thrive in bogs and ponds because of the ability of their roots to withstand waterlogging and to obtain nitrogen from the symbiotic bacterium *Frankia*.

MANGROVES

Of all the habitats that trees have managed to colonize, perhaps the most challenging is the seashore. The mangroves that live around muddy tropical shores have to cope not only with waterlogging but also with the high salinity of sea water, which tends to suck water osmotically from their roots. Mangroves have evolved independently in several different angiosperm families, and they show extensive evolutionary convergence.

The most notable features of mangroves are their roots. The trees are anchored in the soft mud both by prop roots, which emerge from the trunk, and by drop roots, which grow down from their lower branches. Oxygen is piped down through aerenchyma in these roots, and more is also obtained using knee-like pneumatophores. The roots exclude most of the salt from the seawater, but some salt does still enter the tree. To remove it, some mangroves concentrate the salt into old leaves which they then shed. Others, including plants of the genus *Avicennia*, actively excrete salt from salt glands in their leaves. Excluding and removing the salt creates a difficulty for the mangroves, however. Because the water surrounding the roots has a higher osmotic pressure than the water inside, it is much harder for the trees to pull the water into the roots and up to their leaves.

Although they are submerged in water, therefore, mangroves are effectively in permanent drought and have many of the adaptations of desert plants;

they are short, have high density wood, and thick, leathery evergreen leaves that conserve as much of the hard-won water as possible. They also have a reproductive system that is well suited to their shore environment. Many mangroves make use of sea breezes by having wind-pollinated flowers, and the fruits are well adapted for water dispersal; the long, thin fruits of the common *Kandelia* species, for example, can either spear directly into the soil if they are dropped at low tide or float away if they are dropped at high tide. By this stage the fruits have already germinated, so they can rapidly anchor themselves and start growing.

Mangroves readily colonize open areas of mud, and are important in stabilizing and building up tropical shores. Mud builds up around the tangle of roots, and so the water gets shallower. Eventually so much mud builds up around the landward side of the mangrove forest that the land rises above the high-water mark, and the mangroves are replaced by terrestrial trees. Mangroves are also important habitats in their own right, providing a home for a wide range of invertebrates, fish – and even monkeys such as Borneo's most bizarre inhabitant, the proboscis monkey.

One puzzle that remains is why mangroves are restricted to tropical and subtropical areas. The answer lies with their physiology. In temperatures where sea temperatures fall below 20°C (68°F), the cost to the trees of excluding and removing salt seems to be just too great.

BELOW Mangroves in coastal Borneo, showing the prop roots that give them extra stability in the waterlogged soils and pipe oxygen down to the fine roots.

BELOW RIGHT The spear-like fruit of the mangrove *Kandelia*. The long extended organ is actually an enlarged seed root or radicle, which can embed itself into the mud when the fruit falls from the tree.

DESERT TREES

Deserts provide very poor habitats for trees. Their main difficulty is in supplying enough water for transpiration in the face of a hot, desiccating Sun and low and unpredictable rainfall. The few trees that do manage to survive in deserts do so by using two different methods.

MAINTAINING A WATER SUPPLY

One technique of surviving is to reduce water loss as far as possible and to maintain a water supply even during the severest droughts. Many desert trees reduce their water loss by having small leaves with a thick cuticle and sunken stomata, and they protect their leaves from herbivores using thorns. They use a range of techniques to actually obtain their water.

The acacias and tamarisks of Africa and *Prosopis* of Central America are what are known as phreatophytes: plants that obtain water from the permanent water table that lies deep below the soil using tap roots which are up to 50 metres (164 feet) long. This is a very successful technique; the main difficulty such a plant has is in establishing itself, as this would involve its tap root penetrating through metres of bone-dry soil. Phreatophytes are therefore restricted to areas such as

BELOW The grossly swollen trunk of the sack-of-potatoes or Socotran desert rose tree *Adenium obesum* subsp. *sokotranum* from the Indian Ocean island of Socotra, Yemen. The trunk stores water for use during droughts, and offers vital protection against summer monsoonal winds.

ABOVE **The dragon's blood tree,** *Dracaena cinnabari*, **of Socotra, Yemen is a relative of the dragon tree,** *Dracaena draco*, **of the Canary Islands. The reduced foliage and mass of spongy wood helps both trees survive drought. Pollen records indicate that ancestors of these bizarre trees once grew throughout the Mediterranean.**

dried up riverbeds, where the soil is occasionally wetted for long periods of time. Other desert trees obtain water from morning dew, while in the coastal deserts of Chile acacia trees can obtain water by condensing drops of sea fog on their foliage and letting it drop down to their roots.

STORING WATER

The other technique of desert survival is to capture and store water during the occasional rainstorms and use it to photosynthesize during the long droughts. For trees the best place to store water is in a swollen trunk. The famous baobab of the African savannahs is the champion in this respect; a 15-metre (49-foot)-high tree can have a trunk with a diameter of a staggering 9 metres (nearly 30 feet).

In even drier conditions trees exhibit extreme adaptations in their foliage. The dragon trees of North Africa and Joshua trees of North America have strap-like evergreen leaves, which occur only at the tips of their branches. The trees of the genus *Pachypodium*, which grow in the Madagascan deserts, have even fewer leaves, which are deciduous, falling off in the dry season to reduce water loss.

Even then, however, the trees photosynthesize using their bare trunks, which are protected from herbivores by spines. The trend to reduce leaf area is taken to its ultimate limit by the tree cacti of North and South America and the euphorbias of Africa. Members of both families have reduced their leaves to spines. They photosynthesize using only their corrugated trunks, which swell up like bellows after rainstorms. Tree cacti grow slowly, but species such as the giant saguaro can grow to well over 10 metres (33 feet) in height.

ABOVE The giant saguaro cactus, *Carnegiea gigantea*, Saguaro National Park, Tucson, USA, is the ultimate desert tree, totally lacking leaves and photosynthesizing and storing water in its thick corrugated trunk.

CHAPTER 8

Southern hemisphere trees

THERE ARE EXAMPLES OF ALMOST ALL TYPES of climatic zone and hence types of forest in the southern hemisphere. Boreal forest is missing, but that is only because there is no land at latitudes between 55 degrees and the icy wastes of Antarctica. However, although the southern forests are superficially similar to northern forests, their species composition is very different. There are very few of the common northern hemisphere conifers such as yews, cypresses, cedars or redwoods. In particular, none of the most recent genera that are so dominant in the northern hemisphere – the pines, spruces, firs and larches – are present. Most of the important genera of temperate angiosperm trees are also absent south of the central tropics. There are no oaks, beeches, maples, ashes, elms, limes, birches,

Araucaria

Podocarpaceae

Nothofagus

Eucalyptus

LEFT The strange disjunct distributions of some southern hemisphere trees. They only make sense in the light of the isolation and subsequent break-up of the southern supercontinent, Gondwanaland.

OPPOSITE The curtain fig tree of Yungaburra, Queensland, Australia. This is actually a variety of the strangler fig, *Ficus virens*, which is found through tropical Asia and Australasia, and which grows up and around host trees, eventually killing them. The host of this tree has actually tilted over, hence the spectacular curtain of aerial roots.

poplars or hickories (to name but a few!). Instead, the southern hemisphere is filled with what seems to be a relict flora consisting of primitive conifers such as the araucarias and podocarps, and angiosperms such as eucalypts and southern beeches. Many of the groups have very odd 'disjunct' distributions, just like those of the marsupial mammals; they are often found as far afield as South America and Australasia but nowhere in between.

PLATE TECTONICS

So why do northern and southern forests have such different floras? The answer lies in the history of the Earth, and in particular in plate tectonics, the movement of the continental plates that has been occurring since the supercontinent Pangaea started to break up, around 170 million years ago.

The first event was the splitting of Pangaea into two: a northern part, Laurasia, made up of North America, Europe and Asia, and a southern part, Gondwanaland, made up of present-day South America, Africa, India, Antarctica and Australasia. Gondwanaland itself later broke up, starting with the separation of Africa around 120 million years ago, followed by the break-up of Australasia, Antarctica and India. The result was that the southern areas were isolated from the northern ones

BELOW The positions of the continents 170 million years ago. The supercontinent Gondwanaland, composed of South America, Africa, India, Australia and Antarctica is just breaking off the northern supercontinent of Laurasia.

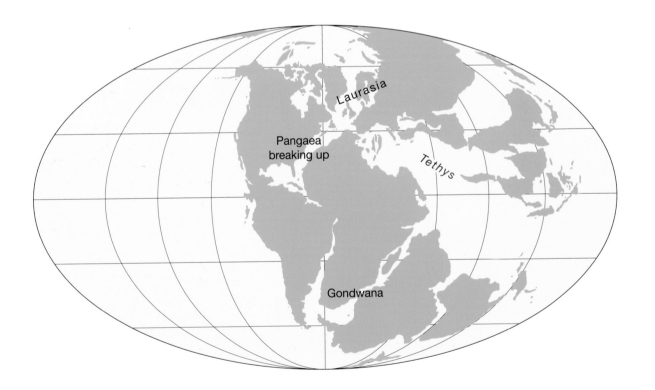

and from each other for many millions of years. It was only 40 million years ago that Africa was reunited with Europe, 20 million years ago that India hit Asia, and some three million years ago that North and South America were rejoined. Australasia still has no land bridge with Southeast Asia, although there is an almost continuous chain of islands, and Antarctica is isolated down at the bottom of the globe.

THE EFFECT ON TREE DISTRIBUTION

What effect did plate tectonics have on the trees? The initial break-up isolated the northern and southern continents soon after the first angiosperms had appeared. Following the break-up, the most rapid tree evolution seems to have taken place in the northern supercontinent, possibly because it was positioned in tropical latitudes and had a more equable climate. The old groups of conifers, such as the araucarias and podocarps, were replaced by the new ones: cypresses, yews and pines. Similarly, southern beeches were replaced by a diverse broadleaved angiosperm flora.

BELOW LEFT *Podocarpus falcatus*, a giant podocarp that lives in the montane forest of southern Africa.

BELOW The Wollemi pine, *Wollemia nobilis*, a relative of the monkey puzzle that was only discovered in 1994 in the Blue Mountains, New South Wales, Australia.

In the southern continents the existing groups of trees such as the podocarps continued to dominate in most areas. Of course, they also continued to evolve, so once the southern continents had separated from each other, their tree floras also started to diverge from each other. Novel groups of conifers that look superficially like cypresses evolved in the three main continents: South America, Africa and Australasia. Africa lacks two important groups that are seen in the rest of the southern hemisphere – araucarias and southern beeches – probably because it was isolated first from the other continents. South America and Australia have a more similar tree flora, since they were joined, via Antarctica, for much longer. Araucarias similar to those that dominated during the age of reptiles are found in both continents and form forests in the warmer, drier areas. The latest discovery of such a tree was made only in 1994, when the Wollemi pine, *Wollemia nobilis*, a relative of the monkey puzzle that was thought to have died out in the Jurassic period, was found in a canyon in the Blue Mountains of New South Wales, Australia. Forests of southern beeches also thrive in the wetter temperate regions of both South America and Australasia.

TREES ON OCEANIC ISLANDS

Oceanic islands have an unusual tree flora, just like the southern continents, and for the same reason – isolation. The fact is that trees are very poor colonizers of islands; their seeds are too heavy to be blown to an island, and they are seldom able to survive a voyage by sea. Two exceptions are the seeds of mangroves and coconut palms; the latter produce the familiar seeds that are buoyed up and protected from saltwater by a huge corky fruit. They are such successful colonizers that the beaches of tropical islands are invariably fringed by coconut palms. Apart from mangroves and coconuts, however, different types of tropical island often have quite different vegetation depending on how they were formed.

Many oceanic islands, such as New Zealand, Madagascar and the Seychelles, are actually made of small pieces of continental shelf that have split off from a major continent. They tend to have a relict tree flora that is similar to that of the continent from which they split. Perhaps the most fascinating island is New Caledonia, which lies several hundred kilometres/miles off the east coast of Australia. It is home to a stunning collection of ancient and bizarre trees: southern beeches, a monkey puzzle, podocarps and many other unusual conifers. It even has the only parasitic conifer, *Parasitaxus usta*! So primeval is the habitat on the island that it provided an ideal backdrop to the antics of the computer-generated dinosaurs in the television series Walking with Dinosaurs.

Other oceanic islands were formed more recently by undersea volcanic action and so have never been attached to the mainland. As we have seen, few trees are capable of colonizing oceanic islands, so most of their trees must have evolved there. A good example can be seen on the Hawaiian Islands in the central

Pacific. These islands were colonized a few million years ago by a daisy-like plant from California called a tarweed. Since then the plant has evolved to produce 28 different species, with a whole range of body plans. Not only are there small herbs like the original plant, but also shrubs and lianas, and even trees. *Dubautia reticulata* is a tree that can reach a height of over 8 metres (26 feet) and can have a trunk with a diameter of nearly 50 centimetres (20 inches).

ABOVE South coast pines, *Araucaria columnaris*, growing on the isolated island of New Caledonia. Of the 44 species of gymnosperms on the island, 43 are endemic.

EUCALYPTS

The big difference between South America and Australia is the dominance in Australia of the eucalypts. There are some 500 species of the single genus *Eucalyptus,* which appear to have evolved since Australia was isolated. In the absence of many of the other groups of angiosperms, eucalypts have become incredibly successful in Australia, just as the marsupial mammals have done. They dominate the woodland from the tropical rainforests of Queensland down to the cool temperate regions of southern Tasmania. Many species are particularly adept at coping with the droughts and fires to which Australia is subject. They survive fire by sprouting from resting buds that are held beneath the thick bark. Some smaller species of trees have a different strategy. These trees, which are collectively called mallees, have an unusual form. Each tree has several trunks that sprout

from a massive 'lignotuber'. This lignotuber is held partly beneath the soil surface and is very resistant to fire. It has recently been suggested that the dominance of eucalypts in Australia is partly due to the use of fire by the Aborigines, who arrived there some 50,000 years ago.

RECENT MIGRATIONS INTO THE SOUTHERN HEMISPHERE

The final influences on the tree flora of the southern hemisphere have occurred since the southern and northern continents started to reunite. Some northern trees have moved into the southern hemisphere: hollies have invaded South America, Africa and Australia, willows and sumacs have invaded South America and Africa, and alders have invaded South America. The amount of movement has not been particularly great, however. In particular, it seems odd that so few trees with a northern ancestry have managed to migrate into southern Africa, despite the fact that they have had 40 million years to do so. The reason may be that their southward march has been blocked by the Sahara and Kalahari deserts. In South America, in contrast, trees could migrate down the well-watered mountains of the Andes.

OPPOSITE **A plantation of red gum,** *Eucalyptus camaldulensis,* **showing the typical pale, peeling bark of the group.**

ABOVE *Eucalyptus* **trees growing with the mallee form. The multiple trunks sprout from an underground lignotuber that can survive forest fires.**

CHAPTER 9
Trees and people

OUR ANCESTORS CAME DOWN FROM the trees and left the forest around five million years ago. Yet the story of humans remains inextricably bound up with that of trees. This final chapter first of all examines how and why we find wood and other products of trees so useful, and goes on to investigate how we exploit trees. It ends with a history of humanity, looking at how our evolution and the development of our civilizations have been shaped by our crucial and continuing relationship with trees.

WHY WOOD IS SO USEFUL

The reasons that we find wood so useful spring directly from the ways in which it has evolved to become useful to the trees themselves. As we saw in an earlier chapter, wood is a superb structural material that is ideally designed to hold up the leaves against the forces of gravity and the wind. With its combination of stiffness, strength, toughness and lightness, this also makes it ideal for us!

In fact, using wood is not quite so simple as felling a tree and cutting it into the right shapes, because we do not use timber in its original 'green' state. We first dry it out. This improves it in two ways. First, dried, 'seasoned' wood is much lighter

OPPOSITE Heddal Stave church, the largest of the Norwegian, stave churches is a masterpiece of wooden architecture. It was built in the 13th century entirely of wood, from the structural beams and planks walls to the split-wood shingles that tile the roof.

LEFT Wood is readily dried, or 'seasoned' by exposing it to the air. The water escapes both from the exposed ends of the trunk and from the sides, if the bark is removed.

because it contains less water; second, drying also makes it two to three times as stiff as wood in its green state. This makes it ideal for our technology, which tends to combine rigid elements with more mobile joints. Fortunately, wood dries fairly rapidly, a fact related to another of its functions (wood cells are tubes that are designed to transport water, and therefore it evaporates readily out of the cut ends). The relatively rapid drying of wood also makes it ideal for its second main use: as a fuel. Wood can be harvested and, if cut up into small pieces, can be dried and burned efficiently after only a small delay.

CARPENTRY AND MODIFIED WOOD PRODUCTS

The only problem with using wood as a structural material is the fact that it is much stronger across the grain than along it. As we have seen, this does not matter for the tree, because all the forces tend to be parallel to the long axes of the trunk and branches. However, for us this can cause problems. This is why carpentry is such a skilled business; carpenters have to be very careful to prevent wood splitting along the grain, both when they are shaping it and when they are making joints between the cut pieces. The design of the finished article also has to be such that it is stressed only in the right directions. Fortunately, even this problem has recently been overcome, with the development in the last century of plywood. In this material, thin veneers of wood are stuck together using waterproof glue, with the grain of consecutive layers oriented at right angles to each other. This produces a material which is equally strong in all directions, but that retains the high strength and toughness of wood. Nowadays, some 10 per cent of all wood is turned into plywood, and it is much easier for the amateur to use than natural wood.

Smaller offcuts of wood are also shredded and glued together in random arrays to form hardboard, chipboard or fibreboard. These materials are cheap and, like plywood, are easy to shape because they have no grain. However, they are by no means as strong, because the board can break between the individual fibres.

RIGHT Boatbuilding in the Amazon. The carpenters have ensured that the grain is oriented more-or-less along the spars.

THE USES OF DIFFERENT WOODS

We saw earlier in this book that trees with different life histories tend to lay down very different wood. As a consequence, this gives each type of wood its own characteristic properties, which suit it for particular purposes.

WOOD FROM CANOPY AND EMERGENT TREES

Canopy trees, such as the oaks and beeches of deciduous forests, the spruces and redwoods of coniferous forests, and the mahoganies, teaks and meranti trees of tropical rainforests, have large trunks made out of wood that has a medium density and that is impregnated with relatively large amounts of dark-staining defence chemicals. This makes them ideal to use for structural timbers; the wood is strong and long-lasting. However, the durability of the timber depends very much on where the trees come from; trees growing in warmer climates tend to produce greater quantities of more toxic chemicals to defend themselves from pathogens. Consequently, more toxins remain to protect the wood even after it is dead, and this makes the wood more durable. This is why in the age of wooden ships, the Spanish galleons made from tropical mahogany lasted many years longer than the British oak-timbered ships. This is also why dark tropical timber is so heavily prized. Even among temperate trees there is a noticeable difference in the colour and durability of timber; cedar from the Mediterranean region lasts better out of doors than oak, while the northern spruces and pines rot very readily.

Fortunately, the durability of timbers can be improved by coating them with waterproof material, which is why softwood window frames and doors are usually painted or varnished. These days, timber durability can also be improved by pressure impregnating it with creosote or other preservatives; this enables us to use more timber from shorter-lived faster-growing trees even though it is naturally less durable.

Emergent trees, and species that tend to grow in more open situations, such as ash and willow, often have wood that is highly flexible and resilient. This enables the trees to withstand gusty winds, but also means that the timber is suitable for certain specialist uses. The resilience of ash makes it ideal for tool handles and baseball bats, while in Britain one species of willow is specifically cultivated to produce cricket bats.

BELOW The modern replica of Captain Cooke's ship HMS *Endeavour*. Unlike the original, which was probably constructed mainly of oak, it is made of the long-lasting Australian hardwood jarrah and has masts made of whole trunks of Douglas fir.

WOOD FROM UNDERSTORY TREES

The understory trees of the world's forests tend to be smaller, slower-growing and longer-lived than the canopy species. As a consequence, they tend to produce denser wood, which is impregnated with even higher concentrations of defence compounds. The high density makes the timber stronger, harder and more resistant to splitting than wood from canopy trees. This makes it easier to carve and to turn. It also has a more attractive appearance; not only is it darker but because the wood cells are smaller and the growth rings are closer together, it has a finer grain. For these reasons the woods of many understory trees are very valuable and have many specialist uses. Spindle trees are so called because their wood was used in the 19th century to make the hard-wearing spools onto which the cotton thread was spun. Woods such as box (boxwood), cherry and ebony are highly prized for the manufacture of furniture and musical instruments. High density is particularly important for woodwind instruments, as dense wood vibrates more at higher harmonics, giving the instrument a brighter tone.

Understory trees from tropical rainforests often produce the densest, longest-lasting and hardest-wearing woods. Greenheart from South America is used to make lock gates and piers because of its exceptional resistance to rotting. Another tropical wood, lignum vitae, is used to make the bearings of ships' propellers; not only does its hardness make it extremely wear-resistant but, because of the resin it contains, it is self-lubricating.

WOOD FROM PIONEER TREES

As wood factories, the fast-growing but short-lived pioneers are the most productive of all trees. However, as we have seen, the wood they produce has a very low density and is very poorly defended against pathogens. Consequently, the wood is weak, has a wide grain and is highly susceptible to decay. These properties render it unsuitable for use as structural or decorative timbers, and so it tends to be used only to make items that have a limited life. Poplar wood is used for the production of safety matches, while some fast-growing pines are used to make the boards of pallets. The lightest timber of all, balsa wood, is used in some situations where its low density is crucial; for example it provides the 'filling' inbetween aluminium sheets in the manufacture of laminates for aircraft construction. Because it is so full of air it is also used as insulation in gas bulk carrier ships.

By far the most important use of fast-growing tree species, such as pines, birches and eucalyptus, is in the production of paper. The wood fibres are first separated from each other and then softened using powerful chemicals to dissolve away the lignin. A pulp is then produced in which the fibres are arranged randomly, and this is pressed to make a flat but very strong tissue. Finally, this tissue is coated with china clay to make an impermeable surface on which the ink, when applied, will not run.

NON-WOOD PRODUCTS

Wood is not the only useful product that can be obtained from trees; they may also be grown (or simply exploited) for the chemicals they produce, for their leaves, flowers, or most frequently their fruit.

CHEMICALS FROM TREES

Most of the chemicals we exploit from trees are ones the tree has produced to defend itself from pathogens or herbivores. Tannins have been extracted for thousands of years from deciduous trees such as oak and chestnut and used to cure leather. Mediterranean conifers are even better sources for such chemicals, as they pipe large quantities of them to diseased or damaged areas of wood along their resin canals. In the 16th to 19th centuries, many important preservatives were obtained from pines and used in the construction of longer-lasting ships. In particular, maritime pine was the source for such important products as turpentine, pitch and tar.

A rather more serendipitous discovery was that of rubber, which is made from a latex taken by tapping the trunk of the rubber tree *Hevea brasiliensis*. The liquid latex plays an important part in protecting the tree, as it can surround a damaged area and seal it off as it dries. The fact that the dry rubber has excellent mechanical properties, however, is rather fortuitous. Another latex that has turned out to have useful properties is chicle, the basis of chewing gum, which is obtained by tapping the chicozapote tree in the rainforests of Guatemala.

ABOVE **A rubber tree, *Hevea brasiliensis*, being tapped for latex. Latex is actually a milky sap that in life protects the trees, like other members of the spurge family, from pests and diseases. The fact that it can be converted into a useful material is a happy accident.**

FOOD AND DRUGS FROM TREES

Wood and bark are so tough and full of defensive chemicals that they are seldom edible. An exception is the bark of the cinnamon tree, which is an important spice. Extracts from the wood, bark and leaves of other tropical forest plants have also given rise to important medicinal products such as quinine and poisonous alkaloids such as curare and strychnine. In temperate regions, the active product of aspirin, salicylic acid, was first extracted from willow bark, while extracts from the needles of yews have been found to have powerful anti-cancer properties.

Tree leaves are heavily defended against herbivores, so we seldom find them digestible. One of the few types of leaves we do use are those of tea bushes. As well as containing caffeine, a stimulant, tea is also packed with pathogenic chemicals. It is thought that these chemicals help prevent water-borne diseases; this may have helped the tea-drinking populations of Japan and Great Britain to live in large cities even before they had developed effective sanitation.

The flowers of trees also tend to be well defended from herbivores, and so we seldom eat them. The clove, however, has flower buds that have long been used as a spice, and this species was, and indeed still is, an important source of wealth in Southeast Asia. Rather than eat flowers, we often make perfumes from them; in doing this we are exploiting the chemicals that the tree makes to attract pollinating insects. Frangipani and gardenia are just two examples of trees whose flowers are used in the fragrance industry.

The fruits of many trees are designed to be good to eat, so that the tree can get its seed transported, and therefore it is no surprise that tree fruits are commonly exploited for food. There are a bewildering variety of tropical fruits, while in regions with a Mediterranean climate there are groves of citrus fruit, olives and almonds. In northern Europe and the northeast USA most of the important fruit trees are members of the rose family. Apples, pears, plums and cherries are grown in huge orchards. The trend nowadays is to grow dwarfed trees. This makes them easier to harvest, of course, but it also has advantages that relate to the ways trees grow. The smaller trees produce proportionally more fruit because less energy is diverted into producing wood, and there is a much shorter trunk up which water has to be transported.

Tree seeds often contain large energy stores that power the early growth of the seedling. However, this makes them attractive to herbivores, so they are often heavily defended by the tree, using a range of chemicals. Many seeds are poisonous, but the less toxic ones are exploited by humans, since the chemicals can give them a strong flavour. They include spices such as nutmeg, and coffee and cocoa tree seeds, which are used in drinks. Other species produce edible nuts that are defended mechanically, using hard shells. We exploit many of these species commercially and grow them in

ABOVE A grove of olive trees, *Olea europaea*, on a terraced hillside, Majorca. Olives are cultivated throughout the Mediterranean for their oil-bearing fruit.

RIGHT A breadfruit tree, *Artocarpus altilis*. Originally from the South Pacific, where its huge fruit are used to make flour, it was transported by the British to the Caribbean, most notoriously on HMS *Bounty*.

orchards; Brazil nuts, however, cannot be cultivated because the plantations lack the necessary pollinating insects, and they have instead to be collected from the wild in the forests of Amazonia. Another tropical tree exploited for its seeds is the oil palm. It is grown in huge plantations, and various parts of its kernel are used to produce edible and industrial oils.

HOW WE EXPLOIT TREES

Forests that have been untouched by humans – so-called old growth or primary forest – generally contain superb trees that have grown to their full potential. Because the trees are sheltered by their neighbours until they reach the canopy, they grow extremely tall, with a straight trunk, and they branch only near the top. This means they contain large amounts of good-quality timber with straight grain, and few of the knots that are formed where branches emerge from the trunk. It is not surprising, therefore, that primary forests are so often exploited. Tropical rainforests have until recently remained the largest areas of pristine unexploited forest, but even they have been subjected to heavy logging since the Second World War, as mechanization has made timber extraction easier.

LOGGING

The main problem with our exploitation of primary forest is that it is so often done in an unsustainable way. Obviously clear-felling an area of forest means that it cannot produce more timber for a long period of time. Indeed, the forest may never recover its vitality because of soil erosion or because of the inability of some of the species to recolonize the area. However, even if we are more careful and just

LEFT Logging operations in rainforests are extremely destructive processes, as can be seen here in northern Perak, Malaysia.

remove the more valuable trees, we will still tend to degrade the forest; we will reduce the proportion of the more useful species and might damage the soil or some of the remaining trees. Many forests in Southeast Asia have recently been subjected to 'selective logging' in which only large specimens of valuable timber species are cut down. Unfortunately, the use of heavy machinery and the speed of the logging operations means that the result is often still devastation; felling the timber trees damages most of the other trees because they are joined together by lianas; heavy extraction equipment destroys many of the sapling trees and compacts much of the topsoil; and bare soil is leached away by the rain. Areas of forest damaged in this way may take centuries to recover naturally.

Fortunately, we long ago developed ways of managing forests that are essentially sustainable; the challenge for the future is to modify them to suit the modern mechanized world, and to create economic and political conditions in which they can be used.

TRADITIONAL SILVICULTURE

One method of sustainably maintaining a forest that has been carried out in northern Europe for hundreds of years is known as silviculture. Trees are extracted one at a time or in small clearings and are replaced naturally by saplings that spring up in the light gap. This technique works extremely well, producing trees that have grown naturally; being sheltered and growing up towards the light, they develop into fine, straight trees with good-quality timber. There is no need in this system to plant young trees; the forester just has to select the best specimens of the most economically valuable species. This technique can, over the centuries, dramatically alter the species composition of a forest without it looking noticeably

RIGHT Broadleaved woodland such as this can be managed sustainably by removing individual trees, leaving a canopy to shelter new regrowth.

'unnatural'. The forests of southeast England, for instance, were dominated some four or five thousand years ago by lime trees (basswoods). These grow and coppice well and so were used extensively for hundreds of years despite the fact that they produce rather poor-quality timber. However, as Britain rose to prominence as a naval power, limes started gradually to be replaced by oaks and beeches, which could be used for shipbuilding. Silvicultural techniques were also successfully used by British forest officers in Asian colonies in the late-19th and early-20th centuries; here, valuable trees with dense wood were extracted individually using elephants.

The main disadvantage with using silvicultural techniques is that they are labour-intensive and unsuitable for large-scale operations. Nevertheless, some effort is being made to employ a similar technique called 'enrichment planting' in logged rainforests. Useful timber species are micropropagated, grown in nurseries, and planted in specially cleared rows in the forest. Young 'weed' species are removed to prevent the timber trees from becoming engulfed by lianas or shaded by other trees.

COPPICING

Coppicing is a variation on the silvicultural management techniques that was developed in northern Europe and was common in the Middle Ages. This technique makes use of the fact that many species of angiosperm trees, such as hazel (hazelnut), chestnut, oak and ash, do not die when they are cut down. Instead they sprout from buds around the base of the trunk to produce large numbers of narrow poles. Growth of these poles is very rapid because the tree already has a fully developed root system to supply it with water and nutrients, and there is no trunk to use up energy or reduce the water supply to the leaves. Coppicing therefore promotes the production of large numbers of narrow woody stems that can be harvested every 10 years or so; they are used to make poles, hurdles, tool handles or other small implements, or they may be burned as firewood or to make charcoal. Many British woods used to be single-species stands of coppiced trees that were developed for small-scale industry. The chestnut coppices of southeast England and France, for instance, were used to make charcoal for brewing and iron-smelting.

In many other woods, however, the coppiced trees were in between a few large 'standard' trees that were grown for timber. The typical English oak/hazel (hazelnut) wood consisted of tall oaks, which were used to make timbers for houses and ships, and coppiced hazels, whose wood had a multitude of uses. An added attraction of these woods was that they often produced a fine crop of acorns and hazelnuts in autumn, which could be used to fatten pigs. In western Scotland, many woods are filled entirely with oaks, the trees with the best form being allowed to grow as standards and poorly shaped ones coppiced.

As labour costs rose in the 20th century and as cheap metal and plastic implements were introduced, coppicing became uneconomic and many former coppices were abandoned. Nowadays they can be recognized only because the coppiced trees have multiple trunks. Recently, however, coppicing has made something of a come-back in woodlands owned by conservation bodies, because it promotes high species diversity. In any one wood only a small proportion of the coppice is cut each year, so the woodland is a patchwork of coppice of different ages; the habitat will range from newly harvested areas, which form an ideal habitat for light-loving flowers and butterflies, to heavily shaded areas, where more typical woodland vegetation is favoured. The idea of coppicing has also been adapted in recent years as a system to produce biofuels. Fast-growing willow and poplar cultivars have been developed that can be coppiced every two years, and their wood is then burned in special power stations.

POLLARDING

A variation on coppicing is pollarding, which involves cutting the trunk and lower branches a metre or two (several feet) off the ground. Though more difficult to do, this promotes the growth of shoots in the same way as coppicing; however, in this case the shoots are sited well away from browsing animals and flood waters. One sort of tree that has a long history of being pollarded is the willow. Pollarded willows are grown widely around marshland areas of Britain and other parts of northern Europe; they produce fine shoots called withies that have been traditionally woven into baskets. Pollarded limes (basswoods) and planes (sycamores) are commonly grown in cities and along roads in continental Europe, as they are less likely than intact trees to be blown down.

BELOW Riverside willows, such as these, are often managed by pollarding, cutting regrowth at a height of around 2 metres (6 feet) to protect the new shoots from browsing herbivores.

PLANTATION FORESTRY

The form of tree cultivation with which we are mostly familiar – plantation forestry – only started to be widely used from the 18th century. In this system trees are planted close together, usually in single-species stands, before being thinned after a number of years. The thinning makes room for the trees that are healthiest and have the best form to mature. They are finally all clear-felled when they reach 'maturity', well before they would actually stop growing. Plantation forestry has many advantages, including simplifying care of the trees and reducing labour costs because heavy machinery can be used for harvesting operations. There are some disadvantages as well, however. For a start, clear-felling can result in the soil being damaged and eroding away. Second, because the young trees are not sheltered by mature ones, they can easily lose their leading shoot, and so develop misshapen stems as a side branch takes over. Single-species stands are also particularly vulnerable to disease, just like a farmer's crop. Finally, of course, single-species plantations by their very nature have a very low diversity. Not only is there only a single tree species, but the understory also tends to be very sparse and to contain only a limited number of species.

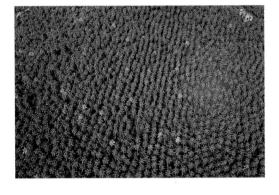

ABOVE A coconut plantation from the air, Papua New Guinea. Coconut and oil palm plantations produce a high yield of seeds but harbour very low biodiversity.

Nowadays foresters are experimenting with techniques to reduce these disadvantages. They are growing trees in mixed-age and mixed-species stands. It is not accidental that these produce growing conditions for the trees that more closely resemble those of natural forests. Indeed, in central Europe, where this approach has been pioneered, it is known as 'back to nature' forestry.

LEFT Harvesting of *Eucalyptus* trees, Eastern Brazil. Modern harvesting machinery can cut, strip and debark whole tree trunks, and even calculate the volume of wood harvested.

TREES AND THE HUMAN STORY

Nowadays humans are terrestrial animals. Many of us live far from any trees: in tundra, in deserts, on grassland plains, or increasingly in cities. To many people, therefore, trees seem to be no more than a curiosity or even a nuisance, dropping sticky sap onto their cars or delaying their train journeys by dropping their leaves. However, closer investigation reveals just how important trees once were and continue to be in our lives – so much so that they have shaped the story of humankind.

OUR ARBOREAL INHERITANCE

In recent years it has started to become clear that our success as terrestrial animals has in many ways been due to the arboreal inheritance we share with our closest relatives – the great apes. Perhaps the most important thing we inherited from them is our high intelligence. Like us, the great apes have larger brains than other mammals of the same size, they are able to make and use tools, and they even show some aspects of self-awareness, being able to recognize themselves in mirrors. It is usually assumed that great apes evolved their intelligence because of their need to communicate within the large groups in which they live, or to help remember the locations of the fruiting trees on which they feed. However, many species of monkeys also live in large groups and feed on much the same food as the great apes, while one of the great apes, the orang-utan, is actually a solitary creature.

The one big difference between great apes and monkeys is in fact their size. Apes can reach weights of 50–100 kilograms (110–220 pounds), which makes living and moving around the canopies of trees extremely hazardous; the branches will bend much further, or even break, under their weight compared with that of a monkey, and a fall is more likely to injure or even kill them. It has therefore been suggested that apes might have developed their larger brains to help them plan safe routes through the forest, while their self-awareness would help them predict how branches would deflect under their weight. This hypothesis is certainly supported by observations of the ways in which the most arboreal of the apes, orang-utans, travel through the forest. They move slowly and carefully, spreading their weight between several branches when they reach the fine twigs at the edge of a tree's canopy. They even use the flexibility of the trees in an ingenious manner to help them travel from tree to tree. Individuals have been seen climbing high in the canopy to cause the trunk to bend over, and using a pumping action to make their tree sway back and forth so that they can reach and move into the branches of the neighbouring tree.

It has even been suggested that one of the most important of our human characteristics – bipedality (walking on two legs) – evolved when our ancestors were still in the trees. Today orang-utans walk upright just like us when they are on narrow branches, which allows them to grip other branches with their hands, both to provide balance and to spread their weight; our ancestors might well have done the same.

But our most obvious inheritance from the great apes is the ability to work wood. All the great apes fashion simple wooden tools, such as fishing sticks, while savannah chimpanzees have recently even been seen to make spears. They break off small branches and strip them of leaves, before sharpening the narrow end into a tip with their teeth. They then use these tools to impale bushbabies that are hiding in hollow tree trunks and drag them out to eat.

However, the most sophisticated woodworking exploits of great apes are seen in their nest-building. All the great apes build sleeping nests at least once a day, something they achieve in an astonishingly short time – less than five minutes – and they seem to know a lot about the properties of wood. Sitting on a solid branch, an orang-utan first constructs the main nest structure; it bends thick branches towards itself, causing them to half break in greenstick fracture, before weaving them together into the floor and rim of the nest. Once a safe structure has been produced, it then reaches up towards thinner twigs, half-breaking them and then twisting them off completely using both hands, before stuffing them beneath itself to make a comfortable mattress. The engineering capabilities of these apes is impressive, and it does not come easily; youngsters spend several years observing their mothers' efforts and building practice nests themselves before they master the craft.

WOOD USE BY EARLY HUMANS

Over the past 15 million years, the climate of Africa has been getting drier. The rainforests have been contracting, and semi-wooded savannahs have been expanding. Early humans moved into this new habitat, which offered different food sources, such as tubers and bulbs and the possibility of hunting or scavenging for meat in the form of large herbivores. While no longer the centre of our ancestors' lives, the savannah trees would have retained their importance as an additional source of food, shelter and safety. The early humans must have inherited the

ABOVE LEFT Orangutans from Southeast Asia are the most arboreal of all the great apes. Despite their great size they live, feed and even sleep in the forest canopy.

ABOVE RIGHT The nest of an orangutan after it has left in the morning. Orangutans build a new nest every evening, bending thicker branches and breaking off thinner twigs to fashion a robust but comfortable sleeping platform.

RIGHT Semi-wooded savannah on the Guyana-Brazil border. Early humans must have evolved in habitats like this in East Africa, having to reduce their dependence on trees.

woodworking skills of the great apes, and probably made wooden shelters and used wooden tools such as spears to hunt animals, and digging sticks to unearth tubers. Unfortunately, though, because of the perishable nature of wood, none of these early tools have survived; archaeologists have instead always concentrated on those tools that have – stone axes. The importance of wood for early humans has therefore generally been greatly underestimated.

Later on, by around half a million years ago, our ancestors had also started to use trees in another way, for fire. In the savannah it is easier to set fire to wood than in the rainforests, because the climate is drier. These fires could then be used to keep people warm through the cold savannah nights, to protect them against predatory carnivores and to cook their food. The first evidence of human controlled fires include charred fragments of mammal bones and plant ash.

The earliest surviving wooden tools that have been found date back even less than half a million years, having been miraculously preserved in the wet acidic soils of Northern Europe. The most impressive finds were the 400,000-year-old spears from Schoningen, Germany: two-metre (6.6-foot)-long throwing spears carved from single spruce stems and grooved at the tip, possibly to hold flint blades. The 'Clacton Spear' – actually only a pointed fragment of yew, is probably of the tip of a spear and was found from even earlier deposits, 450,000 years old in Essex, UK. Recent experimental archaeological tests have suggested that the tip was probably shaped by holding it in a fire and scraping off layers of charred wood with flint blades.

Over the upper Palaeolithic period, wooden tools were progressively developed to increase their efficiency. Spears were hafted with stone tips to improve their

BELOW The broken off tip of the Clacton spear, which dates from around 450 thousand years ago, is the earliest known wooden tool. It was probably weakened during manufacture when the tip was charred by fire to help sharpen it.

penetration, while by 30,000 years ago spear throwers, or atlatls, had been developed to enable smaller dart-like spears to be thrown much further. The development of the bow and arrow (see box) which followed around 10,000 years ago at the start of the Mesolithic period, increased the range projectiles could be propelled still further, and with it improved hunting efficiency.

THE DEVELOPMENT OF WOODWORKING

The emergence of the first farming civilizations, from around 10,000 years ago, required the development of a new range of tools that could help people clear land. At the start of the Neolithic, efficient woodworking axes and adzes (tools with the blade oriented at right angles to the handle) were developed. They had smoothly ground blades hafted into wooden handles, allowing them to be swung with a good deal of energy and to cut deep into wood. Together with wooden and bone wedges, small bone saws and bone or beaver-tooth chisels, they made up a toolkit that enabled large trees to be felled, split into beams and planks, and accurately smoothed and notched. For the first time humans could make large, sophisticated and jointed structures such as houses, dugout canoes and ploughs. By the end of the Neolithic, the first wheels had even been made, carved from a single broad plank, or made by joining together three narrow planks with mortices and dowels.

Woodworking was greatly facilitated by the discovery of metals, first of bronze and then of iron. These metals could be forged or beaten into axes and adzes that had much more slender blades and sharper edges than stone blades and that could fell trees much more efficiently. They could also be incorporated into slender knives and chisels that could be used to make much finer, more accurate joints. These developments greatly helped to improve the transport technology. Wheeled vehicles became much more common in the Bronze Age, and wooden ships such as the North Ferriby boats from Yorkshire, England, could be made watertight for the first time by abutting planks together in precisely fitting joints. Land and ocean-going transport were now much more practicable. The first metal saws were developed by the ancient Egyptians, but it was not until the end of the Middle Ages that saws were actually widely used to cut down trees or saw planks; carpenters continued to process tree trunks by the far less energy-intensive processes of splitting them into planks and then further shaping them using tools that shaved wood off along the grain: adzes, drawknives, spokeshaves and planes.

WOOD IN THE PRE-MODERN AGE

Carpentry did not change greatly after the Iron Age, and wood continued to be the most important material right up to the Industrial Revolution, something that becomes obvious if you visit an open-air museum. Most houses were made of wood, or at least were half timbered, with a framework of wooden beams, and almost all of them had wooden roof trusses. The wall panels, furniture, barrels and other storage vessels, plates and bowls, vehicles, boats and ships were all wooden. Even the engines of the pre-industrial age, windmills and water mills, had wooden sails and gears. Finally, wood was the most important fuel for cooking and keeping people warm.

All this wood had to sourced, however, from a shrinking resource of woodland, because much of the forest had been cleared for agriculture. Many of the civilizations of the near east destroyed their forests, causing erosion and soil loss, helping to lead to their own demise. Human beings had also deforested much of Europe even before historic times; in Great Britain, for instance, which was once covered in forest, tree cover had fallen to below 25 per cent as early as the end of the Neolithic, and now has a tree cover of only around 10 per cent. Much of the east coast of North America was also cleared of forest, first by native peoples, and then, to a greater extent, by the early European settlers.

THE YEW LONGBOW

The English yew longbow is perhaps the most perfect example of the clever use of wood. Yew is a highly resilient wood that is capable of storing a large amount of energy and releasing it into the arrow. The design of the bow uses these properties to their fullest extent. A potential problem with wooden bows is that when you draw them the wood is bent. This causes the front edge to be stretched, while the rear edge is compressed. No single region of the trunk is good at resisting both forces; sapwood is good at resisting stretching, but buckles when compressed; heartwood, which is packed with resins, is good at resisting compression but cracks easily when stretched.

The solution developed by bowyers was to cut the bow from the join between the sapwood and heartwood. The front edge of the bow is made from sapwood, while the rear edge is made from heartwood. This perfectly matches the wood's properties to its function. The front of the bow is stretched when the bow is drawn back, and this is resisted strongly by the sapwood. The rear of the bow is put into compression, and this is strongly resisted by the heartwood. The result is a superb weapon, capable of storing over 100 joules (24 calories) of energy: enough to propel arrows over 100 metres (328 feet) and through the thickest body armour.

Being such an important commodity, it's not surprising, therefore, that control of wood supplies has long been a major factor in power politics. Ancient Egypt had a constant struggle with its neighbours, for example, to maintain its access to a supply of wood from Phoenicia, especially the famed Cedars of Lebanon. Similarly, the need of the British to maintain their wood supply led them to expand their navy partly to keep control of the Baltic timber trade. So important was the need for large tree trunks to make masts that the British even imposed unwelcome restrictions on the use of the huge white pine trees that grew in its North American colonies – restrictions that were widely resented and that helped lead to the outbreak of the American War of Independence.

As the Industrial Age approached there were even bigger demands on the supply of wood, because it was increasingly needed as an industrial fuel and, in the form of charcoal, as a smelting material. Wood availability therefore soon became a limiting factor preventing the expansion of the production of iron, glass, pottery and gunpowder, amongst other important commodities.

TREES IN THE INDUSTRIAL AGE

Perhaps the most important developments that launched the Industrial Revolution were the replacement of wood as a fuel by coal, and of charcoal as a smelting material by coke. These changes loosened the limits of production and allowed rapidly accelerating industrialization and mechanization. Wood did not immediately lose its importance. The first mass production machinery, for instance, was actually produced to make wooden items; Mark Isambard Brunel, the father of the more famous Isambard Kingdom Brunel, designed a series of machines to manufacture the huge number of pulleys that were needed by the British navy. However, as a structural material, wood was not well suited to the demands of mass production. As we have seen it is not only highly variable, but also has very different properties in different directions; it therefore performs much better when wooden items are shaped individually by craftsmen. Moreover, wood is not as stiff or as heat-resistant as metals are. Therefore, the Industrial Age saw wood gradually being replaced by iron, steel, aluminium, and finally plastics for most structural and technological purposes. There was just one final, brief, flowering of wood as a high-tech structural material, in the construction of early aeroplanes, where its light weight overcame its other disadvantages, but after the First World War many woodlands ceased to be actively managed.

Despite its technological demise, though, timber remains a most important material for building; almost all houses still have wooden roof trusses and floorboards, for instance. And timber has been joined as a material by a range of other recently developed modified wood materials, such as plywood, hardboard and

ABOVE Windmills such as this postmill from Sussex, England, represent the epitome of pre-industrial wooden technology. Almost everything, from the building and sails to the cogs and shaft are made of wood.

fibreboard. These new materials are much better suited to our mass production age, and are widely used for furniture making, so we still have plenty of wood about us!

Moreover, just as the technological importance of wood started to wane during the industrial revolution, trees started to become more important than ever to people in two major ways. First, with the rise of literacy in the 19th century, the consumption of books and newspapers expanded rapidly. Up until this time, paper had been made using textile fibres, but the supply of old rags that was needed started to become insufficient. Industrialists therefore developed new ways of mechanically and chemically treating wood to produce bleached wood fibres that could be pressed into paper. Today, almost all paper is made from pulped wood.

URBAN TREES

Perhaps the most significant development was a new environmental use of trees: to improve our towns and cities. Early towns were small crowded areas, often constrained within protective walls. They had narrow streets and small gardens, but this did not greatly matter to their inhabitants, who were well within walking distance of the surrounding countryside. This all changed with the rapid expansion of cities in the Industrial Revolution, and the spread of smoke-

RIGHT **Parkroyal Hotel in Singapore is a spectacular example of the modern use of urban trees. They make the building more sustainable by providing shade and reducing runoff of rainwater.**

producing mills. The workers became crowded into slums, well away from the fresh air and peace of the countryside. Philanthropists among the manufacturing class were moved to address this problem and bring trees to the people. The first urban park in Britain was Derby Arboretum in Derby, paid for by the industrialist Joseph Strutt and landscaped by John Claudius Loudon. Other cities quickly followed, and Joseph Paxton's publicly funded Birkenhead Park, opened in 1847 near Liverpool, was the model for the most famous urban park of all: New York's Central Park. Designed by Calvert Vaux and by Frederick Law Olmsted, who had visited Birkenhead, Central Park was opened in 1858. Trees also started to be planted in the streets and boulevards of major cities, the species chosen being ones that were resistant to the high levels of sulphur dioxide produced by the coal fires of the day: the famous London planes and Manchester poplars.

ABOVE An infrared photo of a city block showing the cooling effect of trees (blue), which can be over 20°C cooler than the surrounding buildings and roads (orange and red).

Not only are urban trees pleasant to look at, but recent research has shown that they also have important environmental benefits. The process of urbanization results in the replacement of vegetation by buildings and roads, and this makes cities hotter in summer and more prone to surface flooding. Urban trees reverse these effects. The water loss from their leaves provides cooling, just as we cool our bodies down by sweating. Trees can remove over half of the energy from the Sun that falls on their leaves in this way, so that even small street trees can provide as much cooling as two air conditioning units. By losing water from their leaves, trees therefore effectively cool the area around them, so that city parks may be a couple of degrees cooler than the surrounding streets. But the biggest effect of trees on their surroundings is due to the shade they provide; the tarmac beneath a tree can be 20–25°C (68–77°F) cooler than tarmac in the Sun, and we can actually feel 10–15° cooler. It's no surprise, then, that people flock to the shady boulevards of Paris and plazas of Mediterranean towns. Research in the USA has also shown that tree shading can reduce the air conditioning costs of buildings by 30 per cent.

Urban trees also help prevent flooding by intercepting rain in their canopies, and allowing water to seep into the soil beneath them. Studies suggest that half the rain falling on a tree's canopy can be diverted from the drains in this way, but the effectiveness of trees can be increased by incorporating them into sustainable urban drainage systems (SUDS). In these systems, water draining from building and roads is diverted into planted areas, which filter it, greatly reducing the volume of runoff and the pollutant load it holds. With the greater prevalence of heat waves and size of storms that climate change is bringing, urban trees are becoming more important than ever in keeping our cities inhabitable.

Urban trees are also good for both our physical and mental health. Their leaves capture pollution particles with diameters of the order of 10 micrometres (0.4 thousandths of an inch) that are emitted by diesel engines, and that cause bronchial problems, lowering pollution levels by as much as 20 per cent in some cities. There is also excellent evidence that trees also make city-dwellers feel happier. Exercising among, and working with, trees improves people's mental state, and even living in buildings surrounded by trees improves our well-being.

THE FUTURE OF TREES

Despite the eclipse of their economic importance since the Industrial Revolution, so great has been population growth that trees continue to be overexploited. Trees are cut down not only for their wood; much forest in South America is being permanently destroyed by being converted into farmland – often for crops such as soya that are fed to cattle and other animals rather than used to feed people. In Southeast Asia, rainforest is also being destroyed to make way for the planting of oil palms, ironically to produce 'green' biofuels. Of the forest that remains, much has been degraded by exploiting it for commercial timber, while on a smaller scale, forests are damaged by cutting down trees for firewood or by slash and burn cultivation.

With the continued logging of old growth forests, not only in the developing world, but more alarmingly in the USA, Canada and Australia, the future for trees may at first sight look bleak. Of the 80,000 or so tree species, some 8,000, or 10 per cent, are endangered. Outside national parks and nature reserves there may soon be few areas of primary forest left. Furthermore, in many areas of the world deforestation has started to have terrible consequences, not just locally but regionally, and even globally. As the Amazonian rainforest shrinks there is a smaller area of actively transpiring vegetation; fewer clouds are being formed and rainfall is reduced. The cloud forests of Central America are already suffering. Conversely, deforestation in the Himalayas has led to catastrophic floods in Bangladesh, since there are fewer trees to intercept and soak up the water from the monsoon rains. Finally, deforestation contributes about one-fifth of the increase in carbon dioxide that is the major driver of climate change; as the trees that have been cut down are burned or rot, the carbon they previously stored is released into the atmosphere in the form of carbon dioxide.

BELOW Sumatra is thought to be losing its natural forest vegetation (shown in green) faster than anywhere else in Indonesia, due to clearance for agriculture and plantations and logging of ironwood trees.

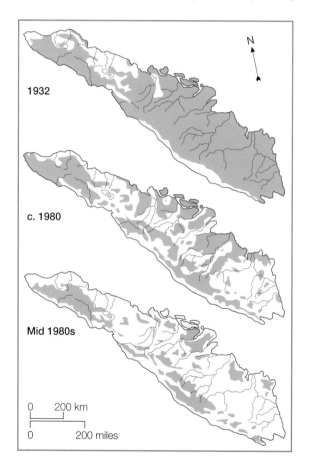

1932

c. 1980

Mid 1980s

0   200 km

0   200 miles

N

CAUSES FOR OPTIMISM

Despite this bleak picture, there are causes for optimism. Many schemes have been set up to measure the ecosystem services provided by existing forests and reward conservation efforts. These include the influential Reducing Emissions from Deforestation and Forest Degradation (UN-REDD) programme for developing countries. There are also many campaigning bodies set up to promote reforestation by both large- and small-scale tree planting, such as the Green Belt Movement, set up in Kenya by 2004 Nobel Peace Prize winner Wangari Maathai. Ecotourism can provide an income stream to local people from conserved forests. Finally, there is growing awareness on the part of both drugs companies and indigenous peoples of the potential for the discovery of useful compounds from forests, emphasizing the value of conserving them. Consequently, rates of forest loss are at last starting to decrease, especially in countries in which governance is improving and prosperity is rising.

A further cause for optimism is growing evidence that even degraded and isolated forests, such as those of Singapore, can maintain the majority of their species. In many parts of the world it is has also been found that forests can recover after deforestation and can once again cover the land. New England, for instance was cleared by the early European colonists for agricultural land that has since been abandoned. It is now covered once again in fine woodland that attracts tourists from all over the world to see its famed autumn colour. Human settlements have also been found in many tropical rainforests that were previously thought to be primary forest, showing that they too can recover after clearance.

We can help, too, by looking after our own woodlands. Sustainable forestry is being developed all over the world, and is being encouraged by such organizations as the Forestry Stewardship Council, based-in Oaxaca, Mexico. There is good evidence from the managed woods of northern Europe that such techniques can keep apparently natural forests going indefinitely.

The final cause for optimism is that forests and trees remain immensely important to people. Trees seem to have a place deep in our subconscious; in many parts of the world, forests and jungles are seen as frightening, untamed places. In contrast, most people's ideal landscape seems to be parkland dotted with trees (probably a similar landscape to the savannahs in which we evolved). Trees are therefore important components of almost all artificial ecosystems: farmland with copses and hedges, golf courses, parks and gardens, and most surprisingly of all even city streets. It seems that our love of trees is hardwired into our systems, thanks to our arboreal inheritance. We must make sure that we never consent to live in a world without trees.

GLOSSARY

Aerenchyma: tissue containing air spaces, which aerate water plants

Angiosperms: flowering plants whose seeds are borne within a mature ovary or fruit

Biomass: total dry weight of organisms in a particular population or area

Boreal forest: evergreen forest found in cool temperate and subarctic conditions

Buttress root: a triangular plate-like root that helps stabilize rainforest trees

Cambium: an area of a plant that produces rows of new cells

Canopy: the upper layer of foliage of a forest

Capillary: a thin tube

Cellulose: a fibrous polymer that is the main constituent of plant cell walls

Chloroplast: a structure within a plant cell where photosynthesis takes place

Climax: the final stage of forest succession

Cohesion: the force that sticks water molecules together

Conifer: a cone-bearing tree

Convergent evolution: the independent evolution of functionally similar structures in distantly related organisms as a result of similar selection pressures

Coppicing: a method of stimulating rapid wood production by cutting the base of the trunk

Cuticle: the waxy outer covering of a land plant, which reduces water loss

Deciduous: a condition in which all the leaves are shed in a particular season

Ecological niche: the role played by a particular organism in its ecosystem

Ectomycorrhiza: a root fungus living as a partner in the roots of a tree, which helps the tree obtain nutrients

Embolism: a break in the water column within a tree

Emergent: a tall tree that extends above the canopy of a forest

Epiphyte: a plant that grows on the outside of another plant

Evergreen: condition in which trees retain leaves all year round

Fibre: a long, narrow wood cell used solely for support

Gymnosperms: seed plants whose seeds are not enclosed in an ovary, the most familiar group being the conifers

Heartwood: the dead wood towards the centre of a tree trunk

Hemicellulose: a polymer that forms the matrix of plant cell walls

Herbaceous plant (herb): a non-woody plant with a relatively short-lived shoot system

Hydraulic resistance: the resistance of water to flow

Lenticel: a pore in the bark of a tree that allows the entry of air

Liana: a woody climbing plant common in rainforests

Lignin: a polymer that strengthens the wall of wood cells

Mutualism: a symbiotic relationship in which both partners benefit

Osmosis: the movement of water across a semi-permeable membrane from a solution with a low concentration of solute to one of high concentration

Ovule: the female spore of a seed plant

Perennating: surviving over a dormant season

Phloem: the sugar-conducting tissue of a plant

Photosynthesis: the biological process in which light energy is used to produce sugars from carbon dioxide and water

Phreatophyte: a plant that obtains water from a deep water table

Pioneer: a tree that is adapted to invade disturbed habitats

Plywood: material made up of layers of wood veneer

Pneumatophore: a knee root of a swamp tree that aerates its root system

Pollarding: a method of promoting rapid wood production by cutting the trunk well above the ground

Pollen grain: the male spore of a seed plant

Pre-stress: internal stress in trees that helps prevent failure

Ray: a radial line of cells in a woody trunk that stores sugars and prevents splitting

Reaction wood: wood laid down by trees to bend their trunk and branches to a new orientation

Rhizoid: underground cells that absorb water and nutrients

Rhizome: a horizontal underground stem

Rhizophore: an anchoring structure in early club mosses

Sapwood: the living wood towards the outside of a trunk that conducts water

Seed: the mature reproductive structure of a seed plant that germinates to form a single offspring

Silviculture: a traditional method of sustainably growing trees for timber

Spore: a specialized reproductive cell that divides to form offspring

Stoma (pl. stomata): minute pore in the skin of land plants that allows gas exchange

Symbiosis: the relationship between two organisms living in close association, which can be harmful to one (parasitism) or beneficial to both (mutualism)

Taiga: northern conifer, or boreal, forest

Thigmomorphogenesis: growth response of plants to wind that helps stabilize them

Tracheid: a long, thick-walled wood cell with a role in both water transport and support

Transpiration: the loss of water from plants

Turgor: internal water pressure within plant cells that helps support them

Understory: the level of vegetation growing beneath the canopy in a forest

Vascular cambium: the ring of cells around a seed plant where new tissues are laid down

Vessel: a wide wood cell used for water conduction in flowering plants

Wood: the water-conducting and strengthening tissue of trees, composed of xylem cells

Xylem: the water-conducting tissue of plants

FURTHER INFORMATION

Selected books

Archibold, O (1995), *Ecology of world vegetation*, Kluwer Academic Publishers, Amsterdam, The Netherlands.

Beerling, D (2008), *The emerald planet: How plants changed Earth's history*, Oxford University Press, Oxford, UK.

Bradshaw, A, Hunt, B and Walmsley, T (1995), *Trees in the urban landscape*, E & FN Spon, London.

Ennos, R and Sheffield, E (2000), *Plant life*, Blackwell Science, Oxford, UK.

Hora, B (ed) (1981), *The Oxford encyclopedia of trees of the world*, Oxford University Press, Oxford, UK.

Horn, H (1974), *The adaptive geometry of trees*, Princeton University Press, Princeton, NJ.

Mattheck, C, (1991), *Trees: Their mechanical design*, Springer-Verlag, Berlin.

Mitchell, A (1974), *A field guide to the trees of Britain and Northern Europe*, Collins, London.

Perlin, J (2005), *A forest journey: The story of wood and civilization*, W. W. Norton & Company, New York.

Rackham, O (2001), *Trees and woodland in the British landscape*, Weidenfeld & Nicolson, London.

Thomas, P (second edition, 2014), *Trees: Their natural history*, Cambridge University Press: Cambridge, UK.

Tudge, C (2006), *The secret life of trees: How they live and why they matter*, Penguin, London.

Whitmore, T (1990), *An introduction to tropical rainforests*, Oxford University Press, Oxford, UK.

Willis, K and McElwain, J (2013), *The evolution of plants*, Oxford University Press, Oxford, UK.

Useful websites

NB website addresses are subject to change.

The Arboricultural Association:
www.trees.org.uk
[an organization to promote the well-being of trees and training of arboriculturalists]

Ancient Tree Forum:
http://www.ancienttreeforum.co.uk/
[an organization championing, mapping and conserving ancient and veteran trees in Britain]

CSIRO:
www.csiro.au/
[Australia's scientific research institution, one area of research being the forestry, wood and paper industries]

The Forestry Commission of Great Britain:
www.forestry.gov.uk
[the UK's forest authority, which manages large areas of accessible forests]

The Forestry Stewardship Council:
www.fscoax.org/index.html
[an organization that promotes the sustainable management of forests throughout the world]

Leafsnap:
http://leafsnap.com/
[electronic field guide to the trees of the USA]

Leafsnap UK app:
http://www.nhm.ac.uk/take-part/identify-nature/leafsnap-uk-app.html
[an app to identify trees from photos of their leaves]

Smithsonian Tropical Research Institute:
http://www.ctfs.si.edu/group/About/
[a network of scientists studying tropical rainforests]

The Tree Council:
www.treecouncil.org.uk
[an organization that encourages tree planting and disseminates knowledge about trees and their management]

Tree-Id – British Tree Identification Guide:
https://itunes.apple.com/gb/app/tree-id-british-tree-identification/id330025326?mt=8
[an app to identify British trees]

UNEP-WCMC's Forest Programme:
http://www.unep-wcmc.org/
[integrated and accessible information on the conservation of the world's forests and their biodiversity]

USDA Forest Service:
www.fs.fed.us
[the USA forest authority, which manages large areas of forests]

Woodland Trust:
www.woodland-trust.org.uk
[the leading UK woodland conservation charity]

WWF:
http://www.wwf.org.uk/what_we_do/forests/
[the forest conservation work of the World Wide Fund for Nature (WWF)]

INDEX

Page numbers in *italic* refer to illustration captions. Where trees are indexed their wood is implied.

abscission layer 76
acacias 91, 92
Acer palmatum 59
Adansonia grandidieri 87
Adenium obesum 91
aerenchyma 87, 88, 89
Aesculus hippocastanum 80
Agathis 72
    australis 20, *20*, 23
ageing 7, 35, 42, 53, 66
air currents and seasonality 69–70, *70*
air quality 122
alders 89, *89*, 101
Alnus glutinosa 89
anchorage 44–6
angiosperms 22
    advances 23
    evolutionary radiation 24
    reaction wood 51, *51*
    temperate broadleaves 80–2
    in tropical rainforests 71–2
    water transport 33–4
ants and rainforest pioneers 74, *75*
apes 114–15, *115*
Araucaria 95
    araucana 20
    columnaris 99
Araucariaceae 20, *96, 97*, 98
Artocarpus altilis 108
ash trees 34, *34*, 43, *105*, 111
Avicennia 89

'back to nature' forestry 113
bald cypresses *see* swamp cypresses
balsa trees 67, 74, 106
bamboos 12, 27, *27*
banana trees 27
baobabs 87, 92
bark 79, 82, 107
    birches 66
    formation of 17
    mechanical design 43
basswoods *see* lime trees
beeches 30, 34, *34, 49, 61, 63,* 65, 80, 82, *105*, 111
    southern 98
Betula pendula 62

bifacial vascular cambium 17
birch sap wine 34
birches 5, 30, 34, *62, 66, 67,* 82, 84, 106
bluebells 81, *81*
bordered pits 33, *33*
boreal forests 22, 82, *83*, 84
box (boxwood) trees 77, 106
branching 17, 19, 44
brazil nut trees 109
breadfruit trees *108*
brittleheart *41*, 42
broadleaved trees 80–2, *83, 110*
bromeliads *73*
buds *80*

cacti 93, *93*
Calamites 13, *13*, 16
cambium 17
canopy trees 64–5, *85*, 105
capillary action 31
carbon sinks 15–16, 122
Carnegiea gigantea 93
carob trees 78
carpentry *see* woodworking
Castanea sativa 48
catkins 80, *80*
cauliflory 72, *72*
Cecropia 12, 38, *38, 74, 74, 75*
cedars 22, 78, 105, 119
cellulose microfibrils 40, 51
charcoal 111, 119
chemicals from trees 107
cherry trees 24, 106
chestnut trees 107, 111
    horse *80*
    sweet *48*, 82
chicle 107
chloroplasts 29
churches, wooden *103*
cinnamon trees 107
Clacton spear 116, *116*
climate
    and plant growth 69–70
    and tree distribution 70–1
climate change 16, 22, 55, 82, 115, 121, 122
climax forest 66
climbers 24, 80
club mosses *11*, 11–12, 15, 16
coal swamps 15–16
coconut plantations *113*
cohesion of water 31–2
cold tolerance 32, 34, 55, 84, 85
compression crease 40

compression wood 51, *51*
conifers 19–20, *20*, 82, 84, 89, 98
    deciduous 77
    reaction wood 51, *51*
    water transport 33
convergent evolution 11, 89
cooling 32, 121, *121*
coppicing 56, *56*, 111–12
cork cambium 17
Cornus nuttallii 65
Corylus avellana 56, *80*
crown *7*, 19, 64, *74*
Cupressus 51
Cyathea 9
cycads 18, *18*–19
cycadeoids 18, *18, 19*
cypresses *51*, 78, *79*, 88, 97

deciduous trees 19, 43, 76, 77–8, 80
defence chemicals 65, 66, 73, 75, 82, 105, 106, 107, 108
deforestation 109–10, 118, *122*
    recovery from 123
dendrochronology 55
desert rose trees *91*
desert trees 26, *91*, 91–3
Dicksonia squarrosa 14
distribution *71*
    limits to 70–1
    southern hemisphere trees 95–101
dogwoods 65, 66, 80, 83
Dracaena
    cinnabari 92
    draco 92
Dracontomelon dao 46
dragon trees 26, 92, *92*
dragon's blood trees 92
drip tip 66, *66*
drought tolerance 23, *26*, 55, 56, 76, 89, 91–3, 99
drugs, medicinal 75, 107, 123
Dubautia reticulata 99

early vascular plants 9–10
ebony trees 66, 106
ecological niches 64–6, 72
ecosystem services 121, 122, 123
embolisms 32, 33, 34, *35*, 77
emergent trees 6, *29*, 64, 105
enrichment planting 111
epiphytes *73, 73*, 77
Equisetum 15, 16

Eucalyptus 42, 57, *95*, 99, 101, 106, *113*
    camaldulensis 101
euphorbias 93
evergreen trees 22, *26*, 72, 77–8, 82, 92
evolution of trees 9–27, 75, 97, 98

Fagus sylvatica 30, 34, 49, 61
ferns *77, 80, 81, 82, 83*
    seed *16*, 17
    tree *9, 14*, 14–15, 17
Ficus virens 95
fig 72
    strangler 73, *95*
fire *117*
    protection from 17, 79
    recolonization after 67, 101, *101*
firs 22, 82, 84
    Douglas *105*
flagging *47*, 47
floods 122
    prevention 121
flowers 24, *24*
food from trees 107–8
forces, resistance to 37–51
forest ecology 67
forks 44
Frankia 89, *89*
Fraxinus excelsior 34
fruit trees 108
fuels 104, 119
future of trees 122–3

Ginkgo biloba 19, 19
glasswort 10
global vegetation distribution *71*
Gondwanaland *95, 96, 96*
grain 38, *39, 48, 49*, 104, *104,* 106, 109
grass trees 25, *25*
Green Belt Movement 123
greenheart 106
Griselinia littoralis 35
growth
    and climate 69–70
growth responses 47–51
growth rings 17, *34*, 39, 106
gymnosperms 17–22

hazel (hazelnut) trees 56, 66, 80, *80*, 111
heartwood 35, *35*, 55, 66, 118
Hedera helix 80

height *11*, 12, 14, *14*, *15*, 27, 29, *54*, *55*, 92, 93, 99
  limitation 10, 48, 53–7, *85*
  tallest living tree 79, *79*
hemlocks 82
Hevea brasiliensis 107, *107*
hickories 67, 96
hollies 43, 53, 77, 101
hornbeams 80
horsetails 12–13, *13*, 15, *15*, 16
humans
  arboreal inheritance 114–15
  early habitats 115, *116*
  suse of wood 44, *103*, 103–8, 115–20
Hyacinthoides non-scripta 81
hydraulic limitation hypothesis 56–7

ironwood trees *122*

jarrah wood *105*
jet 20
Joshua trees 26, *26*, 92

Kandelia 90, *90*
kapong tree *81*
kauri pines 20, *20*
knots 44, 109
Koompassia excelsa 6, *6*, *29*, 64, *64*
'krummholtz' form *47*, *48*

laburnums 24
larches 22, *35*, 77, 84
Larix europea x L. japonica 35
latex 107, *107*
leaves
  arrangement 59–63
  drip tips 72, *72*
  reconfiguration of 43, 64
  see also photosynthesis
lenticels 87, 89
Lepidodendron 11, 12, *12*, 15, 16
lianas 24, 73, *73*, 77, 80, 110
lichens 77, 80, 84
light, competition for 6, 9, 17, 24
  and leaf arrangement 59–63
lignin 10, 40, 106
lignotubers 101, *101*
lignum vitae 106
lime trees 80, 111, 112
locust trees 24
logging 20, 75, 109–10, 122, *122*
longbows 118
longevity 7, 55, 66, 67, 79
lycophytes 11

Maathai, Wangari 123
Magnolia 24
mahogany 65, 105
mallees 99, *101*
mangroves 89–90, *90*, 98
maple syrup 34

maples 30, 34, 43, *59*, 63, 66, 67, 82
maturation hypothesis 55–6
mechanical design 37–51
Mediterranean forests 22, 78–9
Medullosa 16, 17
meranti 65, 105
migrations, southward 101
mistletoe 81, *81*
modified wood products 104, 119–20
monkey puzzle trees 12, 20, *20*, 98
monocot trees 24–7
monolayer trees 59, *60*, 60–3
monsoon forests *76*, 76–7
montane forests 84–5
mosses 11–12, 15, 16, 26, 80, 84
mountain-ashes *see* rowans
multilayer trees *60*, 60–3, 66, 79
mycorrhizas 35

New Zealand privets 35
nitrogen fixation 89, *89*
non-wood products 107–9
Nothofagus 95
number of tree species 5, 59, 71, 122
'nurse trees' *83*
nutrient capture 35, 44
nutrient limitation hypothesis 55

oaks 34, *42*, *45*, 53, *54*, 65, 80, 105, *105*, 107, 111
  cork 43, *43*, 79
  evergreen 77, 78, 79
oceanic islands 98–9
Olea europaea 108
old forests 66, 109, 122
oldest trees 55
olive trees 77, 78, *108*
orangutans 114, *115*
orchards 108

Pachypodium 92
palms 25, 42
  coconut 98, *113*
  oil 109, *113*, 122
Pandanus 25, 26
paper making 106, 120
Parasitaxus usta 98
perfume 108
phenolics 65
phloem 17, 29, 43
photosynthesis 6, 9, 14, 29, *29*, 56, 59–61, 74, 78, *80*, 81, 93, *93*
phreatophytes 91
Phyllostachys nigra 27
Picea sitchensis 33
Pinaceae 22, 82
pines 22, 42, 65, 67, 79, 82, 84, 97, 105, 106
  bristlecone 55, *55*

Calabrian *78*
Canary Island *69*
kauri 20, *20*
limber *47*
maritime 107
New Caledonia 99
Scots (Scotch) 84, *85*
umbrella 78
white 119
Wollemi 97
Pinus
  brutia 78
  canariensis 69
  flexilis 47
  longaeva 55, *55*
  radiata 33
pioneer trees 38, 66–7, *74*, 74–5, 82, 106
pitch 107
plane trees *41*, 112, 121
plantation forestry 113, *113*, *122*
Platanus occidentalis 41
plate tectonics *95*, 96, 97–8, 101
pneumatophores 88, *88*, 89
Podocarpaceae 20, *95*, 96, 97, 98
Podocarpus falcatus 97
pollarding 112, *112*
pollen, evolution of 16, 18
pollination
  by animals 72, 75
  by insects 19, 24, 72, *80*, 108, 109
  by wind 77, 80, 90
poplars 34, 67, 82, 106, 121
  black 88
porosity 34, *34*
Prosopis 91
provenance 85
Psaronius 14, *14*
pterophytes 12, 14

Quercus
  ilex 39
  robur 7, 23, *54*
  suber 43
quillworts 16

radiation of angiosperms 24
rainforests *see* tropical rainforests
rays 23, 29, *34*, 39, *39*
reaction wood 51, *51*
reconfiguration 42–4, 49
red gum trees *101*
redwoods 37, 105
  Californian (coast) 53, 57, *57*, 79
  dawn 22
  giant 5, *57*, 65, 79
reproductive evolution 16, 18, *18*, 19, 23
resins *33*, 35, 106
respiration hypothesis 53

rhizoid cells 9
rhizomes 9, 13, *13*, 27
rhizophores 11
Rhynia 9, 9–10, *10*, 13
riverside trees 87–8
roots
  aerial *14*, 26, 73
  buttress *37*, 46, *46*, 50, *50*, 72
  drop 89
  knee 88, *88*, 89
  lateral 44, *45*, 50, 88
  mechanical design 44–6
  mycorrhizal associations 35
  prop *25*, 89, 90
  seed *90*
  sinker 11, *37*, 44–5, *45*, 46
  tap 44, *45*, 91
rowans 24
rubber trees 107, *107*

sack-of-potatoes trees 91
saguaro 93, *93*
Salicornia 10
salinity 89
Salix
  alba 87
  caprea 80
  fragilis 88
sapwood 35, 118
scalariform plates 34, *35*
Scholander bomb 31
screwpines *25*, 26
secondary succession 67, 75
secondary thickening 17, 19, 24, 26
seed ferns 16, 17
seeds, evolution of 16
Sequoia sempervirens 53, 57, *57*, 79
Sequoiadendron giganteum 57, *65*, 79
shakes in wood 42
shipbuilding 41, *104*, 105, 106, 107, 111, 118, 119
shoot system 6, 9, 12
  mechanical design 42–4
silviculture 110–11
size *see* height
snow, shedding of 44, 84, *84*
southern hemisphere trees 20, 95–101
specialist trees 87–93
spices 107–8
spindle trees 106
spines 93
spiral grain 48
spruces 22, 65, 82, 84, 89, 105, 116
  Sitka *33*, *40*, 53, 85
stomata 9, 29, 56, 91
structure
  trees *7*
  wood 30–1, 38–40

sumacs 101
Sumatra *122*
survival strategies 59–67
sustainable forestry 123
swamp cypresses 22, 77, 88, *88*, 89
swamp trees 88, 88–9
sycamores 67, 82
see also plane trees

tamarisks 91
tannins 107
tar 107
tarweeds 99
Taxaceae 22
Taxodiaceae 22
Taxodium distichum 88
Taxus baccata 22
tea bushes 107
teak 65, 105
temperate areas 22, 77–8, 80–2
tension wood 51, *51*
terpenes 65
Tetrameles nudiflora 81
thickening 17, 19, 24, 25, 26
thigmomorphogenesis 48–50

timber 20, 42, *42*, 65, 103, 105–6, 109–11, 119, 122
tracheids 23, *23*, *30*, 32, 33, *33*, 39
transpiration 29, 31
tree line 48, *85*
Trifolium repens 6
tropical rainforests 6, 15, 46, *46*, 64, *64*, 65, 87, 105, 109, 122
angiosperms 33, 71–2
biodiversity 75
form of trees 72–3
future of 75
logging 109–10
other plants in 24, 77
pioneer trees 74–5
succession in 74–5
trunk
diameter 53, 92
evolution of 10–14, 17
structure 12, *13*, 17, *17*
water storage 92–3, *93*
turgor 9, *10*, 24
turpentine 107

understory trees 65–6, 77, 106
urban trees 50, *120*, 120–2

uses of wood 103–6
early 115–18
pre-modern 118–19
industrial 119–20

vascular tissue 9
vessels 23, *23*, 24, 31–5, *35*
Viscum album 81

water
cohesion 31–2
storage *87*, 92–3, *91*
supply maintenance 76, 91–2
transport 10, *10*, 11, *11*, 29–35, 56–7
waterlogging 87, 88, *89*
Williamsonia 18
willows 5, *80*, 84, 101, 105, 107, 112, *112*
crack 88
wind resistance 37, 42–3
windmills 119
Wollemia nobilis 97, 98
wood
of angiosperms 33–4
of conifers 33

density 74, 106, 111
fracture of 39, *40*
grain 38, 39, *48*, 49, 104, *104*, 106, 109
hydrodynamic design 32–5
mechanical design of 37–42
porosity 34, *34*
pre-stressing 40–2
seasoning 103, *103*
structure 30–1, 38–40
uses of 103–6, 115–20
water transport 10, *10*, 11, *11*, 29–35, 56–7
woodworking 39, 104, 116, 117–18
wound healing *49*, 50, 107

Xanthorrhoea australis 25
xylem 10, *10*, 11, *11*, 17, 23, 29, 56, 77

yews 22, 97, 107, 116, 118
Yucca brevifolia 26

PICTURE CREDITS

p4 Ellen Goff/The Trustees of the Natural History Museum, London; pp 6tl, 14r, 18r, 20l, 21, 22, 25b, 48, 49, 51l, 52, 54, 56, 61, 64, 68, 72r, 75r, 78, 80l and m, 83b, 84, 87, 89, 90 and 108t Roland Ennos; pp 6tr, 27, 97l, 100 and 101 Sean R. Edwards; p7 Adrian Bicker/SPL; p8 Nature Production/NPL; pp 9, 11r, 13l, 14l, 16 and 18l Philip Rye/ The Trustees of the Natural History Museum, London; p10 Prof. William Challoner; p11l, 13r, 17, 29, 37, 39t, 45b, 50, 51r and 70 Mike Eaton/ The Trustees of the Natural History Museum, London; p12 Glasgow Museums; p15 Jos B. Ruiz/NPL; p19, 34, 39b, 45t, 62b and 104 Peter Gasson/Royal Botanic Gardens Kew; pp 20r and 99 Roland Seitre/NPL; pp 23, 30, 33r and 35r Brian Butterfield & Brian Meylan; pp 24, 43l, 66, 74, 80r, 81t and 116t Sandra Knapp; p25t Dr Jeremy Burgess/ SPL; p26 Chris Mattison/NPL; pp 28 and 76 Frans Lanting, Mint Images/SPL; p33l Andre Syred/Microscopix; p35l, 38b and 116b © The Trustees of the Natural History Museum, London; p36 Caroline Jones; p38t Peter Stafford/The Trustees Of The Natural History Museum, London; p40 Prof. George Jeronimidis; pp 41, 63, 75, 85b, 88, 107, 108b and 113t Nancy C. Garwood; p42 Forestry Commission; pp 33l and 43r Andrew Syred/ Microscopix; p46 Fletcher and Bayliss/SPL; p47t and b William D. Bowman; p55 John Shaw/Auscape; pp 57, 79 and 91 Diccon

Alexander; p58 Bjanka Kadic/SPL; pp 60 and 62t Mercer Design/ The Trustees Of the Natural History Museum, London; p65 Floris van Breugel/NPL; pp 71, 95, 96 and 122 Illustrated Image/The Trustees Of The Natural History Museum, London; p72l Wayne Lawler/Auscape; p73l © Patricio Robles Gil/NPL; p81r Forestry Commission; p81b Cameron Hansen/NPL; p83t Doug Wechler/ NPL; p85t Mark Hamblin/TWENTYTWENTYVISION/NPL; p86 Tony Camacho/SPL; p92 Rod Waddington; p93 C. K. Lorenz/SPL; p94 Jonathan Ayres/ The Trustees of the Natural History Museum, London; p97r Jaime Plaza van Roon/Auscape; p102 Destillat/Deposit Photos; p103 Jim West/SPL; pp 105 and 117 Prof Stephen Blackmore; p109 Fletcher and Baylis/SPL; p110 © Bob Gibbons/ardea.com; p112 © Nick Pound; p113b Luiz Claudio Marigo/NPL; p115 Adam can Casteren; p119 Barry Samuels; p120 ©Patrick Bingham-Hall; p121 Sustaining Urban Places Research Lab, Portland State University.

NPL, Nature Photo Library; SPL, Science Photo Library.